Plastics Technology

For Diploma Level Students and Technicians

Plastics Technology

For Diploma Level Students and Technicians

Midhat Luqman
Life-Fellow
The Institution of Engineers (India)
Allahabad, UP

CBS Publishers & Distributors Pvt Ltd

New Delhi • Bengaluru • Chennai • Kochi • Pune
Hyderabad • Kolkata • Mumbai • Nagpur • Patna

Disclaimer

Science and technology are constantly changing fields. New research and experience broaden the scope of information and knowledge. The author has tried his best in giving information available to him while preparing the material for this book. Although, all efforts have been made to ensure optimum accuracy of the material, yet it is quite possible some errors might have been left uncorrected. The publisher, the printer and the author will not be held responsible for any inadvertent errors, omissions or inaccuracies.

Plastics Technology

For Diploma Level Students and Technicians

ISBN: 978-81-239-2266-9

First Edition: 2013

Published by Satish Kumar Jain and produced by Vinod K. Jain for

CBS Publishers & Distributors Pvt Ltd
4819/XI Prahlad Street, 24 Ansari Road, Daryaganj, New Delhi 110 002, India.
Ph: 23289259, 23266861, 23266867 Website: www.cbspd.com
Fax: 011-23243014 e-mail: delhi@cbspd.com; cbspubs@airtelmail.in

Corporate Office: 204 FIE, Industrial Area, Patparganj, Delhi 110 092
Ph: 4934 4934 Fax: 4934 4935 e-mail: publishing@cbspd.com; publicity@cbspd.com

Branches

- **Bengaluru:** Seema House 2975, 17th Cross, K.R. Road, Banasankari 2nd Stage, Bengaluru 560 070, Karnataka
  Ph: +91-80-26771678/79 Fax: +91-80-26771680 e-mail: bangalore@cbspd.com
- **Chennai:** 20, West Park Road, Shenoy Nagar, Chennai 600 030, Tamil Nadu
  Ph: +91-44-26260666, 26208620 Fax: +91-44-42032115 e-mail: chennai@cbspd.com
- **Kochi:** 36/14 Kalluvilakam, Lissie Hospital Road, Kochi 682 018, Kerala
  Ph: +91-484-4059061-65 Fax: +91-484-4059065 e-mail: kochi@cbspd.com
- **Pune:** Bhuruk Prestige, Sr. No. 52/12/2+1+3/2 Narhe, Haveli (Near Katraj-Dehu Road Bypass), Pune 411 041, Maharashtra
  Ph: +91-20-64704058, 64704059, 32342277 Fax: +91-20-24300160 e-mail: pune@cbspd.com

Representatives

- **Hyderabad** 0-9885175004
- **Kolkata** 0-9831437309, 0-9051152362
- **Mumbai** 0-9833017933
- **Nagpur** 0-9021734563
- **Patna** 0-9334159340

Printed at Magic International Pvt Ltd, Greater Noida U.P.

Preface

It is our national policy to stride for becoming a developed nation. This policy is being persuaded since our country became independent. There is a constant and sustained effort to keep the pace of progress in almost every field of human activity. Science and technology is one of them. For sustained and speedy progress in any aspect of development, education of young generation is of utmost importance. For this reason there is a good number of polytechnics all over the country to impart training in various fields of technology. Demand for technically trained diploma holders has profoundly increased due to their abilities on shop floors.

Educating and training our young, talented, innovative and inquisitive minds is the key to cater to their development and progress and the country at large.

Effective training needs good and applicable books for students. Thorough revision and updates are required to refresh the knowledge and applications taught in classrooms and workshops.

Plastics technology is one of the courses which are taught in polytechnics. It covers brief introduction to chemistry of polymers, properties, various production processes, quality aspects, etc. At times, students have to go through many books to cover various topics. Generally, books for specific topics are quite voluminous and contain much more matter than what is needed for diploma students.

This book, *Plastics Technology* is designed and prepared to provide practice oriented knowledge to readers. Effort is made to cover syllabus of plastics technology diploma level courses. Advantage of this effort to students would be that most of the matter, as prescribed in syllabus would be found covered. In case any student or technician wants a detailed and deep study, they can consult other textbooks on particular topic.

At the end of each chapter Test and Assignment are given with the objective to provide readers with the option for self-assessment for knowledge and information assimilation.

Hope this book *Plastics Technology* would be helpful for students to learn and prepare for examinations. Technicians who are already working on shop floor of a plastics factory may find this book useful for finding solution to some of the technical problems or difficulties in their day-to-day working.

Midhat Luqman

Acknowledgements

Writing a book for publication needs good planning and fixing a target time frame for submission of work to the publisher. At the planning stage it was realized that to prepare book *Plastics Technology*, author would need support from his well wishers.

It would be discourteous on my part if I do not mention organizations and my friends who whole-heartedly supported me in preparing the manuscript.

Aurburg, Germany and Windsor India has very kindly provided me with some photographs of their Injection Moulding Machines.

I would like to thank my friend Er PC Joshi for giving me valuable suggestions regarding planning and formatting of book. We discussed book related matters from time-to-time. Thanks to Dr Vjay Kumar, Dy Director CIPET, Lucknow for giving me time and discussed about testing of polymers. My thanks are also due to Prof S Iqbal Ali, Principal University Polytechnic, AMU for allowing me to take photographs of some machines and polymer testing equipment. My ex-colleague Mr Suhail Ansari helped me untiringly in fairing up Figures by AutoCad. Another colleague Mr Jawwad Hasan Khan helped me in Photography.

I have great appreciation for USA based young engineer Mr Moazzamul Azam who time and again helped me designing get up of book and solving problems related to computer operation while preparing text and taking printouts.

In the last but not the least, I want to thanks my wife Nuzhat Luqman who supported and encouraged me during this ordeal, especially in making the monogram effective and impressive. My children also remained a symbol of strength for me during the entire course.

Midhat Luqman

Acknowledgements

Contents

Preface v

1. Plastics Materials and Applications 1

Thermoplastics Materials 1
Thermosetting Materials 18
Description 18
Molding Process 23
Polyblends and Alloys 23
Nanomaterials 26
Test 28
Assignment 28

2. Plastics Processing Techniques 30

Injection Molding 30
Machines 30
Preparation of Plastic Granules and Pellets 47
Feeding of Plastic Granules to Injection Moulding Machine Hopper 49
Plasticization of Granules 50
Injection of Plastics into Mould 52
Flow of Plastics into Sprue, Runner, Gates and Cavities 69
Flow of Polymer Melt into Runner, Gate and Cavity 72
Multicolor Injection Moulding 75
Mould Temperature and Short (Components Plus Runner) and Cooling Time 76
Mould Closing and Opening 79
Moulding Cycle Sequence 81
Ejection System 82
Defect in Moulding, Cause and Remedies 86
Extrusion 94
Thermoforming 109
Blow Moulding 117
Calendaring 122

FRP 125
Plastics Waste Management 129
Test 133
Assignment 135

3. Plastics Testing Techniques 138

Introduction to Plastics Testing 138
ASTM 139
DIN 140
BIS 140
ISO 141
Mechanical Properties 141
Straps 142
Packing Straps 142
Plastics Sheet 142
Travel Suitcase 143
Chairs 143
Small and Big Containers 144
Vehicles Seat Covers and Linings 144
Thermal Properties 144
Performance Properties 145
During Use of Product 147
Utility Time Performances 147
Processing and Flow Properties 148
Processing Temperature 148
Percentage of Fillers 148
Optical Properties 149
Refractive Index 149
Light Transmittance 150
Electrical Properties 150
Dielectric Constant 151
Dielectric Strength 152
Dielectric Dispersion 152
Surface Resistivity 152
Volume Resistivity 153
Product Testing 153
Test 156
Assignment 157

4. Machine Maintenance 159

Maintenance Methods 159
Routine Maintenance 161
Scheduled Maintenance 161
Breakdown Maintenance 163

Hydraulic Principles 164
Principles of Hydraulics 165
Hydraulic Pumps 167
Radial Piston Pumps 169
Axial Piston Pump and Tilted Plate Principle 170
Pressure Relief Valve 170
Directional Control Valve 171
Hydraulic Motors 173
Pressure Control Valve 173
Volume Control Valve 174
Strainers 175
Filters 175
High Pressure Steel Pipes 176
Hoses 176
Joints 176
Electrical Controls 176
Electrical cum Electronic Control of Injection Molding Machine 177
Operation of Circuit 183
Semi-automatic Operation 184
Electronics and Microprocessor Based Components and Equipment 185
PID 186
PLC 186
Management of Plant and Machinery Maintenance 188
Manuals 189
Trained and Dedicated Maintenance Man Power 190
Electronic Technicians 190
History Recording 191
Job Description of Maintenance Fitters 193
Test 195
Assignment 196

5. Quality Management and SQC Techniques 198

Introduction to Quality 198
Process Parameters 200
Smoothness in Motion 201
SQC Techniques and Application 202
Case Study 202
Analysis of Data 206

Concept of Acceptance Sampling 208
ISO 9000, ISO 17025 and ISO 14000 Standards 209
Management of Quality Assurance 210
Preparation of Quality Manual 211
Training of Operators and Workers 211
Display of Instructions 213
Collection of Data 214
Processing the Data on Computer 214
Corrective Measures 214
Regularity in Quality Assurance Related Work 216
Test 216
Assignment 217

6. Costing and Industrial Management 219
Basic Principles of Costing 219
Standard and Marginal Cost 221
Costing of Injection Molded Components 221
Marginal Cost 229
Proforma for Costing 230
Proforma Costing Sheet 230
Introduction to Industrial Management 231
Personality 231
Aptitude 232
Motivation 233
Communication 234
Leadership 235
Organizational Structure 235
Matrix Organizational Structure 239
Test 240
Assignment 240

7. Practice—Plastics Processing 243
Automatic Injection Molding Machine 243
Microprocessor Controlled Injection Molding Machine 243
Extrusion Process 244
Compression and Transfer Molding 244
Automatic Blow Molding Machine 245
Thermoforming Machine 245
How to Run an Injection Molding Machine? 245
Study of Moulds 253
Finishing of Plastics Components 254
Test 255
Assignment 256

8. Practical—Plastics Testing 259

Introduction 259
Various Tests and Testing Equipment 260
  Melt Flow Index 262
  Specific Gravity Measurement 264
  Izod Impact Test 265
  Abrasion Test 266
  Universal Testing Machine 267
  Tensile Strength Test 269
  Coefficient of Friction 270
Testing of Pipes 270
Testing of Water Storage Tank 272
Testing of Films and Sheets 273
Testing of Bottles 276
  Vertical Load Test 276
  Drop Test 277
  Bursting Test 277
Testing of Plastics Products 278
Test Specimen Preparation 279
Test 280
Assignment 281

Index 285

1 Plastics Materials and Applications

THERMOPLASTICS MATERIALS

Plastics materials as, we know and see today, have a long history of development. During Babilonian period, naturally occurring shalac, a family member of plastics, was used for making ornaments and other articles. Reference to rubber is also provided here because it has some relevance to plastics. Around the same time, rubber was also known in its natural form which occurs under the bark of rubber trees which grow abundantly in far east countries and places like Java and Sumatra islands. Rubber in its natural form drips out drop by drop from cuts or broken barks. When exposed to air, it slowly solidifies to become a sticky material.

During last two centuries, lot of discoveries and inventions took place in various fields of technology and science, including molecular science and organic chemistry. Around year 1835, natural raw rubber was treated by adding sulfur and heat under patented process to produce vulcanizing rubber. Patent was granted to Charles Goodyear. Today tyres for automobiles, trucks and two wheelers with trade name *Goodyear* can be seen apart from other makes. In Greek, word polymer means *many parts* and it is used for combination and chains of monomers. Nomenclature monomer and polymer were coined by scientists who worked exhaustively on atoms, molecules, structures of monomer and polymer. Before beginning of 19th century, it was considered by scientists that clusters of molecules are bonded together by some *unknown* force. In 1922, a scientist Hermann Standinger expressed his view that long chains of atoms are held together by covalent bonds. This means that electrons and protons of similar or dissimilar elements hold the atoms together. This is called covalent bond (Fig. 1.1).

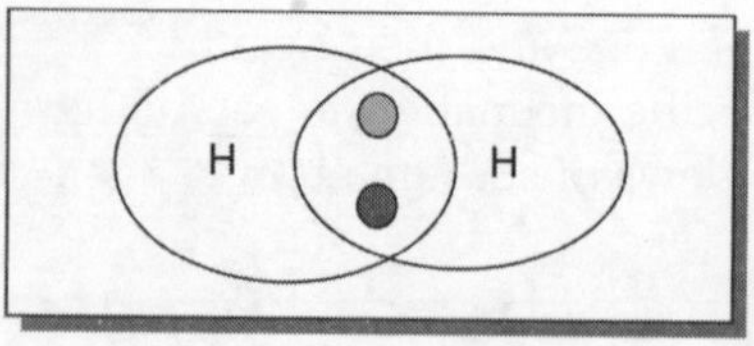

Fig. 1.1: Covalent bond

Basic principle of covalent bond is briefly mentioned in following paragraph to recollect what is taught at school level.

Every substance has a tendency to become stable which they achieve:

- By losing/gaining electrons
- By sharing electrons

To become stable, they must have eight electrons in their last shell (octet structure).

Electrovalency (Transference of electrons)

Metals have 1, 2 or 3 electrons in their outer shell. They can readily lose these to non-metals (which need those) and thus they form a stable electrovalent compound with an electrovalent bond.

Example: Atom of an element is smallest structure on which characteristics and properties of an element depend. Figure 1.2 displays the basic structure of an atom.

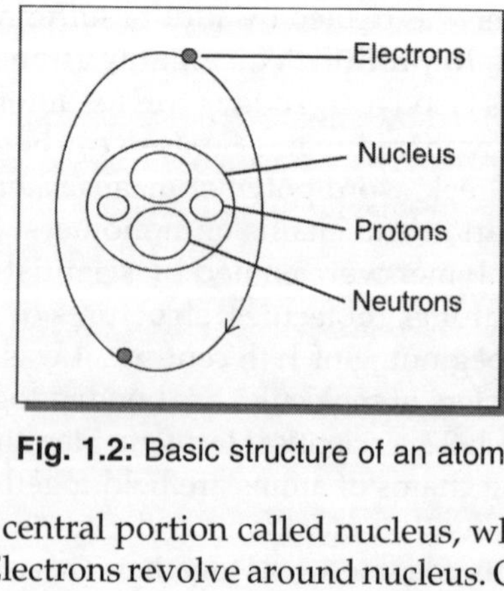

Fig. 1.2: Basic structure of an atom

It consists of central portion called nucleus, which has protons and neutrons. Electrons revolve around nucleus. Orbit of electrons may be more than one and their oval type orbits are in different planes with specific inclination to each other. Figure 1.3 shows representative views of sodium atom.

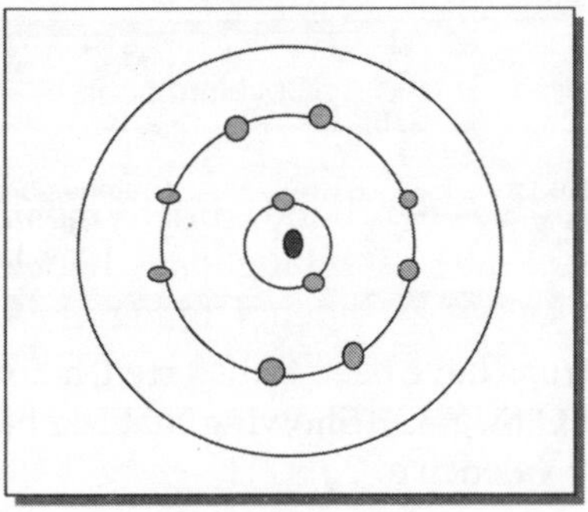

Fig. 1.3: Sodium atom

Please note that view shown in Fig. 1.3 is just a way to explain. Actually, movement of electrons of any element is quite complex and it is very difficult to show on paper.

Atoms are very stable, if outer layer of electron is filled or empty. Filled layer contains 8 electrons with the exception of hydrogen (H) and helium (He), this is called *octet rule*.

Combining of atoms generally takes place in two ways:

1. Gaining or loosing of electrons
2. Sharing of electrons

Compound formed by gaining or loosing is *ionic compound*.

Compound formed by sharing of electrons is *covalent compound*.

Basic concept of atom's bonding is illustrated in Fig. 1.4.

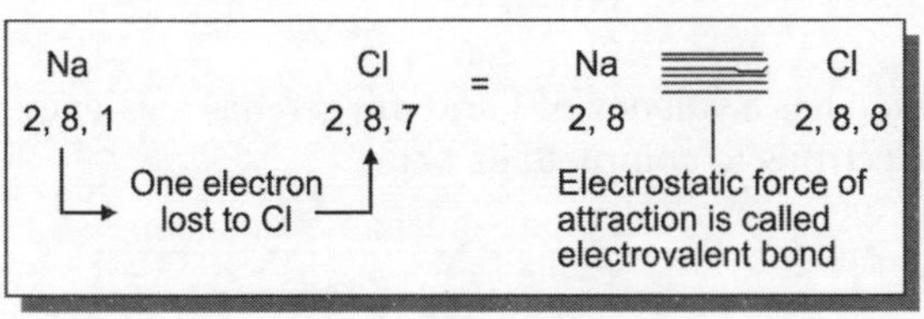

Fig. 1.4: Atom's bond

This also means that sodium is showing a valency of 1 and chlorine also showing a valency of 1. As only 1 electron has been lost/gained, structure may be Na–Cl.

Example

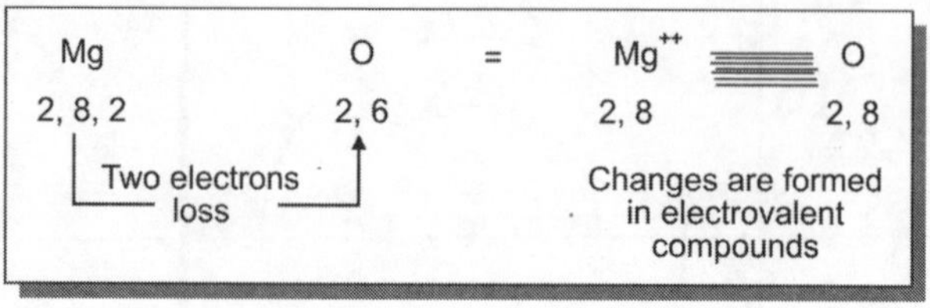

Since two electrons have been transferred, a double bond exists between Mg and O (Mg=O). Showing that Mg has a valency of 2 and oxygen has a valency of 2.

Covalency (Sharing of electrons)

When non-metals combine with each other, they generally do so by mutual sharing of electrons.

Example: Formation of chlorine molecule, sharing configuration is 2, 8, 7.

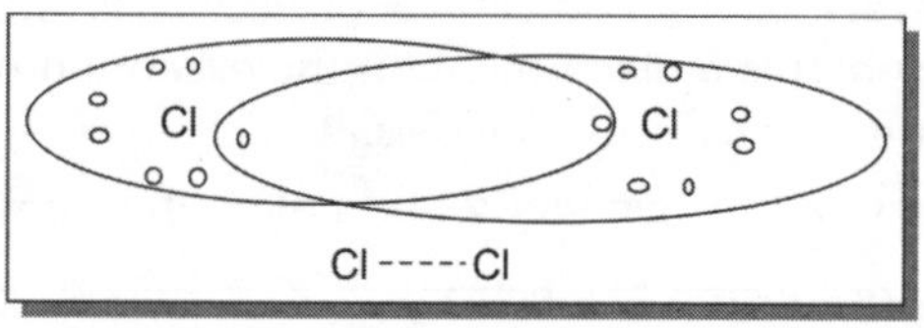

Chlorine is showing a valency of 1. No charged particle is formed. Only last shell electrons are involved.

Formula of H_2O

$$H = 1$$
$$O = 2, 6$$

Hydrogen has a valency of 1 and oxygen has a valency of 2 as it needs 2 electrons to complete its octet.

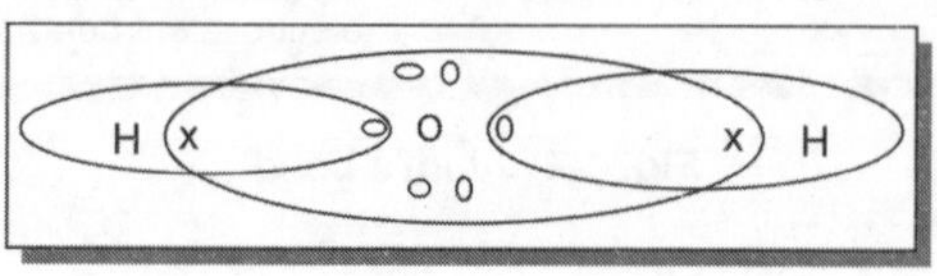

Formula of carbon dioxide

C = 2, 4 (electronic configuration)
O = 2, 6

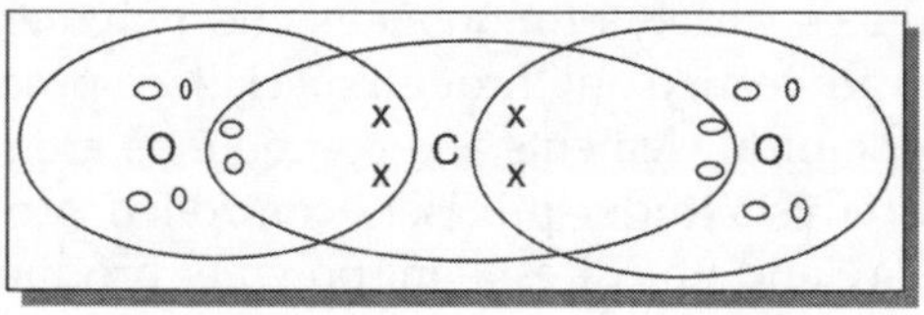

or

O=C=O

Oxygen again showing a valency of 2 and carbon of 4.

Formula of methane: CH_4

C = 2, 4
H = 1

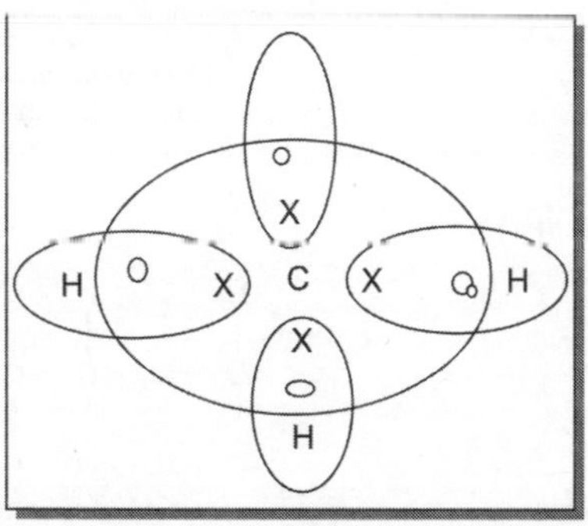

or

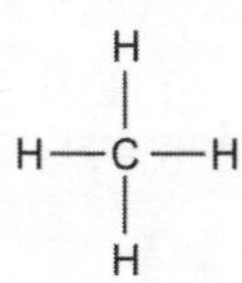

So, carbon again showing a valency of 4 and hydrogen a valency of 1.

Around beginning of 18th century, scientists discovered nitro-cellulose. When it was treated with camphor, it produced celluloid which was probably the first synthetic polymer. Around that time, word plastics was coined for general communication about natural and synthetic monomers, polymers and copolymers.

Chemical combination of various chemical elements and compounds produces variety of monomers and polymers. Chemical combinations are carried out through specialized processes where a variety of chemical elements and compounds are reacted with or without catalysts under precisely controlled process temperature, pressure and time. Resultant polymer products are given various names.

Polyethylene monomers and polymers are chemically represented as shown in Fig. 1.5.

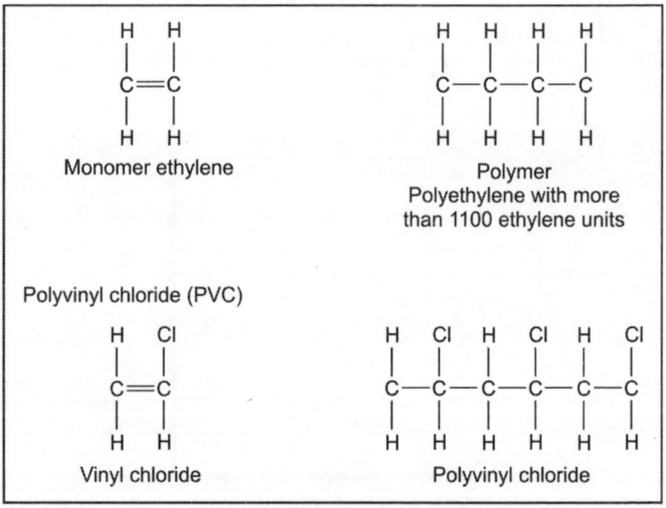

Fig. 1.5: Monomer—polymer

Basically, there are following distinct families of plastics:

- Thermoplastics
- Thermosetting

Thermoplastics are those polymers which can be processed many times for converting into useful products through various production processes such as injection molding, extrusion, blow molding, etc. If a thermoplastics is injection molded, it is subjected to heating. In case, plasticizing temperature exceeds suitable temperature then quality of thermoplastics may some what deteriorate. This may cause further generation of scrap. If generated scrap is stored and handled under very clean condition only then granulated scrap may be reprocessed with slight deterioration in quality of end product. Theoretically, thermoplastics may be processed many times but practically it is not advisable specially for excellent quality products.

Thermoplastics may be manipulated by combining specialized additives, modifiers, plasticizers and by producing copolymers.

There are a number of quality aspects which may be required in end products. Few of them are listed below.

- High temperature functional tolerance
- Low temperature functional tolerance
- Low friction
- Low or no effect of ultraviolet radiation
- Good transmittance of light
- Degradable variety
- Tough, hard, brittle, etc.
- High gloss
- No or little reaction of chemical
- Less or no hygroscopic behavior
- Less or no permeability to gases
- No migration of color pigments, additives and plasticizers

Polymer production processes are design to incorporate any of the above or combination of few requirements. This aspect has motivated plastics producers to develop a variety of types and grades of thermoplastics. Names of some thermoplastics are listed below.

- Polyethylene (PE)
- Polyvinyl chloride (PVC)
- Polypropylene (PP)
- Styrene's (PS)
- Polyamides (PA)
- Polycarbonate (PC)
- Polyethylene terephthalate (PET)
- Acrylonitrile butadine styrene (ABS)
- Polymethyle methacrylate (PMMA)

Most of the above have variety of grades. This makes long list of plastics introduced in market to fulfill the specific needs of consumers.

There are value added polymers with specialized performance as needed by high-tech industry such as aircrafts, space system, satellites, communications, atomic power generation, electronics, medical equipment, etc.

Properties and characteristics of plastics may be classified under following headings:

- Physical
- Mechanical
- Electrical
- Thermal
- General
- Optical
- Flammability
- Glass-like transparency
- Outstanding impact strength
- Very good thermal resistance
- Good electrical properties
- Good dimensional stability
- Rigidity and dimensional stability
- Easy processing
- Good color stability
- Good chemical resistance
- Good abrasion characteristics
- High heat resistance
- Rigidity
- Resistance to weather aging
- Excellent toughness and rigidity
- Excellent abrasion and chemical resistance
- High gloss finish and excellent surface quality

Properties of few polymers are given in Table 1.1 to give an idea of variations in various properties aspect. In fact there are a large number of various types and grades of plastics.

ABS

- Extrusion grade
- Flame-retarded grades of ABS, ABS/PVC, ABS/PC
- ABS/nylon
- ABS/PC injection molding and extrusion
- Injection molding grade heat resistant, medium impact and high impact

- Translucent
- 20% glass fiber reinforced
- 40% long glass fiber reinforced
- EMI shielding (conductive)
- 6% stainless steel fiber
- 10% stainless steel fiber
- 20% carbon fiber
- 20% graphite fiber
- 40% aluminium flake
- Rubber modified

Acetal

- Homopolymer
- Copolymer
- Impact modified homopolymer
- Mineral filled copolymer
- Extrusion and blow molding grade
- Copolymer with 3% silicon-low wear
- UV stabilized copolymer glass reinforced homopolymer (varied percentage)
- PTFE filled homopolymer (varied percentage)

Polyamide

- Nylon type 6 (with a range of glass fiber reinforced)
- Toughened
- High impact rubber modified
- Impact modified
- EMI shielding
- Nylon type 66 (a variety of modification such as flame restarted grade)
- Nylon type 69 (molding and extrusion)
- Nylon type 610 (molding and extrusion compound) glass fiber reinforced
- Flame retardant grade
- Nylon type 612, tough and flame retardant grade
- Nylon type 612 lubricated with 13% PTFE and 29% glass fiber reinforced

Polycarbonate

- High viscosity
- Low viscosity

- Glass reinforced (varying percentage)
- Flame retardant grade
- High heat, polyester copolymer
- Impact modified
- EMI shielding (carbon fiber, graphite fiber) lubricated (varied percentage of silicon or PTFE)

Polyester Thermoplastics

- Polybutylene terephthalate (many varieties)
- Polyester alloy
- Polyethylene terephthalate
- Polyester/polycarbonate blends
- Glass fiber reinforced (a variety of reinforcement)

Polyethylene and Ethylene Copolymers

- Low and medium density
- High density, monomer and copolymer with varied molecular weight
- Cross-linked molding grade
- Cross-linked extrusion grade for wire and cable coverings

(*Reference: Modern Plastics Encyclopedia '98*)

Detailed study can be done after obtaining specification data sheets from producers of various plastics.

Most of plastics manufacturers world over provide values of various properties such as melt flow index, specific gravity, tensile strength, processing temperature, mold shrinkage, etc. Values provided are based on internationally standard test system such as ASTM or BIS. For example, melt flow index is determined by applying D1238 ASTM test method. Student and personnel in industry may study various test methods to have an idea as to what equipment and texting procedure is to be adopted. A separate chapter on quality aspect is provided in this book. Few test methods are briefly described there.

Selection of type and grade of plastics depends on the design of component or article and the service conditions under which it has to perform. Parameters of design are generally decided by mutual discussions between would be producer and designer. Production cost, performance and time frame are main factors which are taken into account by designer, because he has to follow producer's instructions and preferences.

There are innumerable parts and articles which are produced from a number of polymers.

Following is the list of some articles for which choice of plastics may be different.

1. **Household appliances**
    - Dishwasher components
    - Washing machine
    - Refrigerators
    - Freezers
    - Mixers
    - Other utilities
2. **Automobiles**
    - Body seats
    - Toplines
    - Trims
    - Airbag door
    - Bumper
    - Window parts
    - Fuel tank
    - Under the bonnet parts
3. **Electronics**
    - Computer tower
    - Monitor
    - Keyboard
    - Mouse
    - Business machine
    - Meters
    - An array of small components like USBS
    - CDs
    - Electronic typewriter
4. **Aircaft and space vehicles**
    - Outer skins
    - Internal fittings
    - Cockpit screens and fittings
    - Passenger cabin window screens
    - Many other technical and non-technical components

10. Communication

- Exchange instalation gadgets and components
- Cable pipes and coaxial cable
- Telephone instrument
- Mobile bodies
- Mobile accessories

11. Toys for

- Infants
- Children
- Sophisticated toys

Let us take example of few articles and discuss various considerations for the selection of suitable plastics.

Mobile Body

Mobile normally consists of two or three parts. These parts are fitted together with precise tolerance and even click locking notches. Mobile often fell down on hard floors or thrown by children so the plastics should be such that it sustains impact. No damage should occur to any portion of body, cover or latch. Normally, mobiles are not subjected to temperature above 50°C. There are dummy mobiles available at toy shops for children to play with. Components of such mobiles are not accurately fitted and may get damage, if dropped down on hard floor. In case of toy mobile, plastics used may be of cheep varieties. There are many properties of polymers (plastics) which are taken into consideration while selecting type and grade of plastics. Table 1.1 provides comparative properties of three thermoplastics.

There are quite a number of properties, but only few are considered (Table 1.1).

Table 1.1: Selected properties

S.no.	*Property*	*High density*	*High impact*	*ABS*
1.	Toughness	Good	Better	Best
2.	Hardness	Poor	Good	Good
3.	Mold shrinkage range	Variation	Good	Best
4.	Scratch resistance	Poor	Good	Best
5.	Impact	Good	Better	Best
6.	Gloss[1]	Better	Poor	Best

[1] Gloss on a finished component also depends on cavity surface finish.

Cellphone is a value-added high-priced product. Purchaser looks for a good functional, reliable and decent looking product. In spite of the fact that ABS is costlier (as of now) as compared to HDPE and HI polymers, over all effect of ABS cost hardly matters due to advantages of its properties. A toy cellphone has to be non-expensive. There are two options for selection of variety of polymer. HDPE may not be a preferred choice as it does not give a feeling of rigidity. Moreover, visual finish may not be attractive due to possible sink marks. High impact polystyrene may be a preferred choice as outer surface of body and cover may be given a mat surface. Glossy surface cannot be obtained with HIPS. It may be obtained to some extent, if medium impact polystyrene is used.

Next example of an automobile light lens may be considered. Lenses for automobiles and two wheelers can be produce from general purpose polystyrene or acrylic (PMMA). Glass-like clarity is needed which is possible from both polymers. In case of GPP, clarity is there but impact strength is very poor. Consequently, GPP lenses get damaged by slight impact or *hurt*. In contrast, PMMA (polymethyl methacrylate) is a tough plastics with good light transmittance (about 98%) hence it is generally used.

For cheap items, such as small containers are molded out of polystyrene. Photo 1.1 shows few polystyrene components.

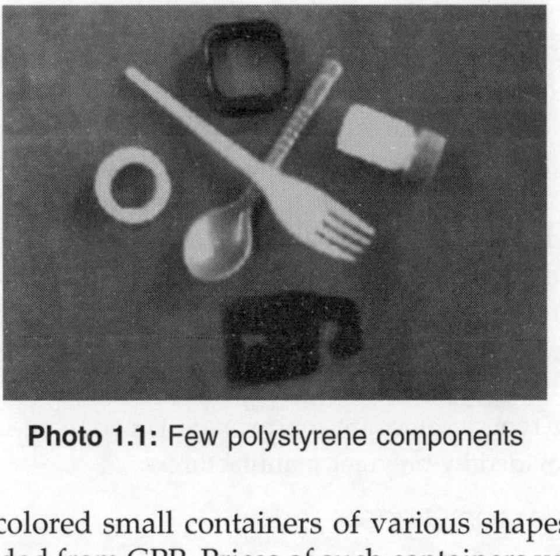

Photo 1.1: Few polystyrene components

Clear or colored small containers of various shapes and sizes may be molded from GPP. Prices of such containers vary in a low price slot.

Sophisticated containers for branded cosmetics and other products are generally fine looking due to plastics quality and finish of molded product. In Photo 1.2, few items are shown to visualize product quality.

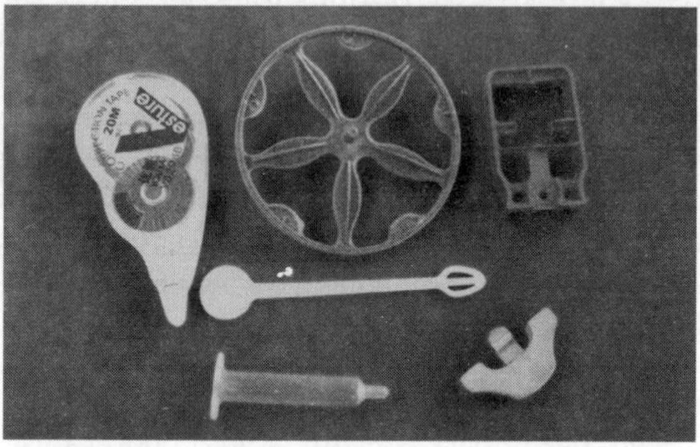

Photo 1.2: Fine polystyrene components

Bottles are generally injection blow molded in plastics such as polycarbonate, polymethyl methacrylate, PET, ABS, etc.

Technical components such as gears, sliding parts, some parts under the bonnet of an automobile are molded in polyamide. Plain polyamide 6.6 or glass reinforced is used. Sections for doors, windows, cabinets and other fittings are generally extruded from copolymer of PVC or high impact polystyrene.

Refer to Table 1.1, properties of three polymers are already described as good, better, best. These are attributes and therefore, do not indicate properties in terms of values obtained by standard tests. Knowledge of specific values of various properties is almost a must for designing mold cavities, runners and gates. It is also necessary for designing extrusion dies and deciding process parameters. Table 1.2 provides information regarding properties, test method and typical values of three polymers.

Note: Given typical values are approximate. For accurate values, consult literature provided by polymer manufacturers.

More Varieties and Grades

Polyethylene generally has three varieties, namely low density, linear low density and high density. There are three important variables

Table 1.2: Properties test methods

S no.	Properties	Test	HDPE	HIP	ABS
1.	Tensile yield strength psi	D 638	2800–3900	2100–8000	5000–7200
2.	Izod impact Ft—lb/in of notch 1/8 inch thick	D 256 A	3.2–4.5	0.95–7.0	3.0–9.6
3.	Mould shrinkage (linear) inch/inch	D 955	0.015–0.040	0.004–0.007	0.004–0.009
4.	Scratch resistance	–	Poor	Good	Good
5.	Melt flow index gm/10 minutes	D 1238	5.4–6.8	5.8	1.1–1.8
6.	Gloss	–	–	–	Excellent
7.	Light transmittance	–	Hazy	–	–
8.	Hardness, Rockwell shore	D785 D2240	Shore D44–50	R50–82; L-60	R102–115

(Reference: S No 1,2,3,5,7,8 extract from modern plastics encyclopedia '98, pages B-192, B-201, B-153)

on which quality depends. These are specific gravity, melt flow index and processing temperatures. Polyethylene can be converted into usable product by processes like injection molding, blow molding, extrusion, thermoforming, rotational molding, etc.

High density polyethylene has a specific gravity around 0.96. It is tough and rigid as compared to low density polyethylene which has specific gravity around 0.93. High density polyethylene retains its rigidness to a great extent even at a temperature of 100°C. Like low density polyethylene, HDPE also has a wax-like feeling, if touched and can be scratched. HDPE producing companies generally produce a number of grades having variation in properties to cater to the needs of processors. Some grades of HDPE are suitable for injection molding process, some for extrusion and some for blow molding. In injection molding variety of HDPE, there are a number of grades having variation in property norms. For example, for molding thin-walled components, a grade with higher melt flow index would suit.

Polypropylene has a specific gravity around 0.91 to 0.92. Hence more numbers of components can be molded as compared to HDPE. Polypropylene(PP) is tough and hard as compared to HDPE. Consequently, it cannot be scratched easily. It is also less flexible as compared to HDPE.

Polystyrene is normally a clear, hard and brittle plastics having a specific gravity around 1.12. It is known as general purpose polystyrene. By adding plasticizers, rubber and modifiers, a high impact grade is produced. It is called high impact polystyrene. With this grade of polystyrene, shiny lustrous surface cannot be achieved in injection molded components and articles. Acrylonitrile butadine styrene (ABS). This variety is considered as engineering plastics by virtue of its properties. It is tough, hard and gives lustrous surface, when molded in a highly polished cavities of injection mold. ABS is available in a number of grades to meet various requirements of processors, namely for extrusion, thermoforming and plating, etc. Normally, ABS is colored during production process to give attractive shades standardized by producers of ABS.

Polymethyl Methacrylate (Acrylic)

It is a rigid and some what tough clear plastics. Its light transmittance is around 98%. Its specific gravity is 1.24 and melt flow index ranges from 1.4 to 27. A number of transparent color shades can be obtained by mixing master batches at mixer-loader of automatic screw-plunger injection molding machines. Depending upon condition of cavity surfaces of injection molds, acrylic (PMMA) provides good surface finish of molded articles. This is also termed as engineering plastics.

Polycarbonate

It is some what rigid and tough clear plastics. Its light transmittance is around 97%. Specific gravity is 1.2 and melt flow index ranges from 3 to 10 gm/10 minutes. A number of transparent color shades can be obtained by mixing master batches at mixer-loader of automatic screw-plunger injection molding machines. Fine filaments can also be extruded from extrusion grades of polycarbonate. It is a clear and tough plastics. It has some advantages in properties over acrylic. It is considered an engineering plastics by virtue of its properties like good mouldability, low percentage of shrinkage at ambient temperature, light stability, good colorability, less migration of pigments. It is also a good plastics for extrusion process.

Polyamides (Nylon)

There are many injection molding grades including glass fiber filled grades. Polyamide 6, 66, 6, 12 and 46 are few grades with different properties. Polyamide is very hygroscopic plastics there-fore, it should be effectively dried before injection molding of a variety of components. Polyamide offers much less friction and much wear resistance in sliding components like bush bearings and linear sliding surfaces in combination of metal and polyamide.

Polyvinyl Chloride (PVC)

This thermoplastics can be modified for a number of properties. It is commonly used for coating electric wires, production of sheets and foils, pipes and pipe fittings and injection molded components for agricultural and electrical gadgets. This plastics is flame retardant and difficult to process because of its sensitivity to heat. It chemically gets degraded and releases hydrochloric acid, when exposed to heat for a longer period of time.

Polyethylene Terephthalate (PET)

This is a clear and tough plastics most suitable for beverages and food products. Jars and bottles can be blow molded by process stretch-blow molding. Suitably dimensioned and shaped preforms are first injection molded and then stretched-blow molded.

Polycarbonate/ABS Blend

Following are some important varieties:

- Flame retardant
- Structural foam
- General purpose
- Glass re-enforced
- Extrusion blow molding

SAN

Following are few grades

- General purpose
- High performance
- High clarity

Polyamide/ABS blend
- General purpose
- Glass re-enforced

Polyester: A variety of grades are available.
Liquid crystal polymer (LCP): This polymer is mostly used for production of electronics components and surgical devices.

Its important properties are:
- Suitable for thin wall molding
- Strong
- Rigid

Dimensional stability can withstand momentary soldering heating.

THERMOSETTING MATERIALS

Description

Thermosetting polymers once heated to curing temperature in a mold and pressed at a pliable stage of curing to compactness in cavity. It is then given time to cure and cool down to a temperature suitable for ejecting the formed component from cavity. Once thermoset material is cured, it cannot be soften by re-heating as is possible with thermoplastics. If cured, thermoset plastics is continuously heated to high temperature then it will decompose and may form ashes. Thermosetting plastics (phenolic resins) are basically of two types— phenol formaldehyde and melamine formaldehyde.

1. Phenol Formaldehyde

- Resole
- Novolac

Resole and novolac are produced by reaction between formaldehyde and phenol but percentage of both chemicals is different which may typically be like below.

Resole ____________	formaldehyde, (excess)	phenol (less)
Novolac ____________	formaldehyde, (less)	phenol (excess)

For combining formaldehyde in less and phenol in excess quantities, require additional catalyst to complete the reaction.

Generally, novolac is in solid state and is mostly used as molding compound. General purpose molding compound typically has following composition:

Phenolics (novolac) ——————40%
Filler (wood flour) ——————60%

Further, phenolic compound used for engineering purpose can be loaded with 70% fillers. These fillers may be combination of short length fiber glass and minerals, carbon, thermoplastic fibers, PTFE and graphite for reducing wear between sliding surfaces.

Phenolics are produced in granular or pellet form. Sheets are generally molded with compound, reinforced with carbon fiber.

There are many varieties and grades of thermosetting materials, few of them are mentioned below.

Epoxy
- Bisphenol molding compound
- Sheet molding compound (SMC)
- Novolac molding compound

Melamine formaldehyde
- Cellulose filled
- Glass fiber reinforced

Phenolic
- Glass fiber reinforced
- Carbon
- Wood flour filled
- Wood flour and mineral filled

Polyester
- Rigid casting thermoset
- Flexible casting thermoset
- Glass fiber reinforced
- Unidirectional
- Premix chopped glass
- Woven cloth
- SMC and BMC low density
- SMC low pressure

EMI (electromagnetic interference) shielding (conductive) SMC, BMC (bulk molding compound)

High strength SMC, 50% glass fiber reinforced

Bismelamides (BMI)

- Produced by the condensation reaction of a diamine with maleic acid.
- Cure temperature 177°C.
- After an elevated post cure temperature (232°C), it exhibits superior properties.
- This polymer is merged into composites as a pre-preg matrix used in electrical printed circuit boards, structural aircraft, aerospace composites, etc.

Epoxy (Epoxide)

Epoxy is widely used in numerous formulation and forms in aircraft and aerospace industry.

Properties of epoxy

- High strength, glass reinforced
- Relative density 1.6 – 2.0
- Thermosetting processing range
  C:300°F–330°F (148.8°C–165.5°C)
  I:280°F–380 °F (137.7°C–193.3°C)
- Molding pressure I–5
- Shrinkage 0.001–0.008
- Tensile strength 5000–20000
- Compressive strength 18000–40000
- Flexural strength 8000–30000
- Linear expansion (10^{-6} in/in/°C) 11–50
- Hardness rockwell M100–112
- Flammability V–0
- Water absorption 24 H(%) 0.04–0.20

Phenolics (PF) and Phenol Formaldehyde Resins

Properties of phenol formaldehyde resin

- High strength glass fiber reinforced
- Relative density 1.6–2.0
- Water absorption 24 H (%) 0.003–102
- Thermoset processing range (°F)
  C: 300–380 (148.8°C–193.3°C)
  I: 330–390 (165.5°C–198.8°C)
- Molding pressure I:20
- Shrinkage 0.001–0.004

• Tensile strength	7000–18000
• Compressive strength	16000–70000
• Flexural strength	12000–60000
• Izod impact (ft–lb/in)	0.5–18.0
• Linear expansion	8–21
• Hardness rockwell	E 54–101
• Flammability	V–0

Polyester

This is an extremely versatile, fairly inexpensive polymer. Unsaturated polyester combines an unsaturated dibasic acid and a glycol dissolved in a monomer, generally styrene, including an inhibitor to stabilize the resin. Organic peroxides, such as methyl ketone peroxide (MEKP) and a promoter are combined with the resin to initiate a room temperature cure. In this liquid state, polyester may be processed by numerous methods, including hand lay up (HLU), vacuum bag molding, spray up and compression molded sheet molding compound (SMC), etc.

The resin can also be B-staged after application to chopped reinforcement and continuous reinforcement, pre-preg. Solid molding compounds in the form of pellets or granules are also used in process, such as compression and transfer molding.

Advantages and disadvantages of thermoset polymer:

Advantages

- Well-established processing and application history
- Overall, better economies than thermoplastic polymer
- Better high temperature properties
- Good wetting and adhesion to reinforcement

Disadvantages

- Resin and composite material must be refrigerated
- Long process cycle
- Reduce impact-toughness
- Poor recycling capabilities
- More difficult repair abilities

2. Melamine Formaldehyde

Melamine formaldehyde is made from the polymerization of formaldehyde (CH_2O) with melamine ($C_3H_6N_6$). Polymerization is chemical process in which two or more identical small molecules called monomers are linked to form a chain of polymer. A polymer is a macromolecule or a large molecule that is comprises bonded monomers.

Chemical formula of melamine formaldehyde is ($C_4H_6N_6O$)

It is white, tasteless, odorless, resistant to chemical, heat resistant, hard, scratch and impact resistant, resistant to shrinkage and heat.

Generally, it is used to make household crockery such as glasses, cups, bowls, plates, pan knobs and handles.

Caution

Kitchen utensils made of melamine should not be used in *micro-wave ovens* as they absorb microwave's radiation and heat. Melamine resin is often used to saturated decorative paper that is laminated under heat and pressure and then pasted on to particle board. The resulting panel is often called melamine and commonly used in ready to assembled furniture and kitchen cabinets.

(Reference: Internet> Wikipedia>Melamine Formaldehyde >Application-In the kitchen)

Parts such as electrical sockets, plugs, switches, contactor body, multi-pin relay base and other items are generally produced from phenol formaldehyde thermosetting material.

Figure 1.6 and Photos 1.3 and 1.4 show few items of thermosetting plastics.

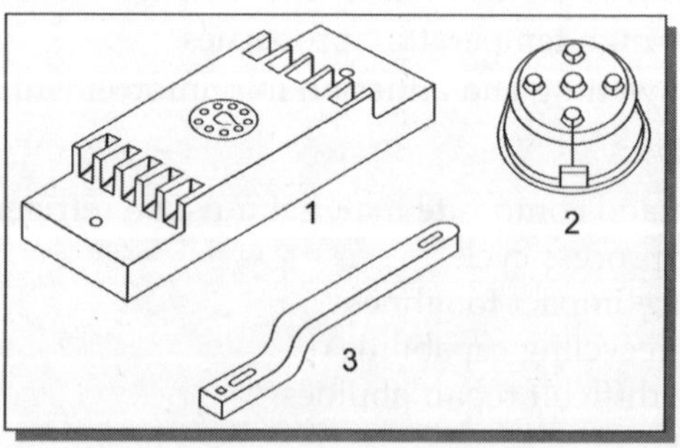

Fig. 1.6: Thermosetting components: Part 1: Relay base; Part 2: Autodistributor cap; Part 3: Pan handle

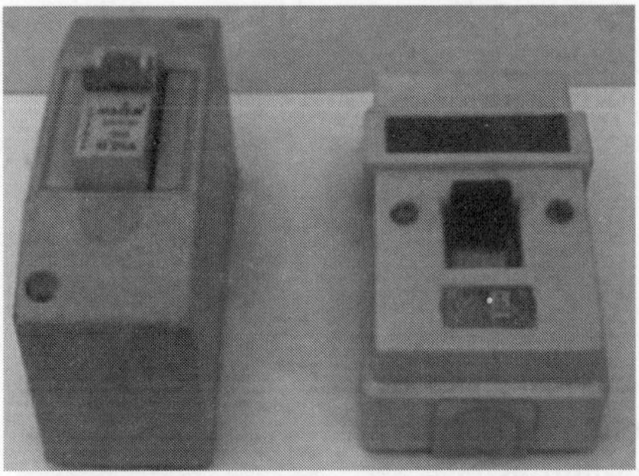

Photo 1.3: Thermosetting components

Photo 1.4: More thermosetting components

Molding Process

Most common process is compression molding. In this process, granules or pallets of thermosetting materials are pre-heated up to temperature, just below temperature on which material starts becoming pliable. This is done to save time in compression molding process and getting a homogeneous pliability. Heated material is then pored into already heated cavity having temperature around 200ºC. In practice, actual temperature is determined by trials.

POLYBLENDS AND ALLOYS

Various plastic materials are modified by mixing modifiers, master batches, filers and even other type and grade of polymers.
Reference: ***Biocomponent Fibre*** *updated April 2004.*

www.engr.utk.edu/mse/pages/textiles/index.html

(On clicking search, page would be open with heading *Material Science* and *Engineering 553* and *554* hence article *Bicomponent Fibres* may be found. Site tried on 20th October 2011.)

Polymers so prepared are given a generic name as *Polyblends*.

There are a variety of additives, few of them are mentioned below.

- Carbon black
- Colorants
- Flame retardants
- Plasticizers
- Fillers, non-fibrous
- Heat stabilizers
- Light stabilizers
- Anti-blocking agents
- Antimicrobials
- Blowing (foaming) agents

Many of them are mixed during production process of polymers. Others are mixed with plastic granules, powder or resin by tumbling or at hopper of machine. Other method is to extrude the mix and form granules.

In case of thermosetting plastic, various types of fillers such as wood flour, chopped fiber-glass, carbon and even aluminum flakes are mixed for compression molding. Such preparation may be termed as polyblend.

In case of thermoplastics, sophisticated method is that two different polymers are extruded from closely spaced die holes, filaments coming out of die holes are continuously spined to form one blend polymer strands/filaments. This type of blending is also done by extruding one polymer as core and other as the surrounding layer of/cover of another polymer. This filament is then spined to form a uniformly blended polymer filament. There is a possibility of having many sectional profiles as shown in Fig. 1.7. This is done to get firm holding of two polymers.

Choice of sectional profile depends on factors such as relative percentages of two polymers, ultimate mechanical properties and mechanical holding together two polymers. Polyblends are also produced by adding additional chemical during polymerization process. In fact, polyblend is a vast subject because of large number

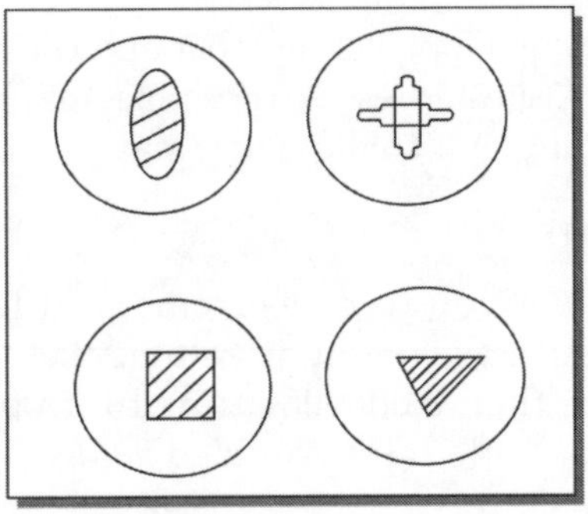

Fig. 1.7: Spin strand profile

of material types, chemical reactions, procedures and methods. It is beyond the scope of this book to go in details.

To further enhance knowledge, it is recommended that textbooks, dealing with 'polyblends' may be studied. Also, on visiting following websites on internet few books may be found.

http://www.thefreelibrary.com

Following books may be located on internet via search engine "Google".

Title: Performance of Multilayer Film Using Maleted Linear Low Density Polyethylene Blend

Authors: sanchez_valdes,s;orona_villarreal,F;lopez_Quinttainill M;Yanez_floresil;devalle.L.F.Ra

Title: The Compatibility of Linear Low Density Polyethylene–Polypropylene

Blends: Viscosity ratio plots

Authors: Bains. M; Balke. S.T; Reck. D; Hon. J

Further, there is a reference for *Bicomponents*. Details may be read from article *Bicomponent Fibers* updated April 2004, Raghavendra R; Hegde; Atul Dahiya; MG. Kamath (Praveen Kumar Jangala and Ramaiah Kotra). This article may be available in Paper 'Material Science and Engineering, 553 and 554. This may be reached by writing following address bar of URL (address bar).

www.engr.utk.edu/mse/pages/Textiles/index.html

Practically, it is found that by writing above website on address bar, desired topic may not be found. Alternatively, write

Bicomponent fibers on Google search bar and click. A page would open having many site addresses. Those sites may be visited just by clicking site heading. A lot of literature may be found.

NANOMATERIALS

Nano is a prefix given to measurement of length or other measurements such as weight or volume. Let the base unit of length be meter, then 0.000000001 or 10^{-9}th part of meter is prefixed as nano. Table 1.3 is provided below to give an idea of various prefixes.

Table 1.3: Prefix

Prefix	*Measurement*	*Scientific notation*
Kilo	1000 m	1×10^{3}
Hecta	100 m	1×10^{2}
Deca	100 m	1×10^{1}
Base	1 m	1×10^{0}
Deci	0.1 m	1×10^{-1}
Centi	0.01 m	1×10^{-2}
Milli	0.001 m	1×10^{-3}
Micro	0.00001 m	1×10^{-6}
Nano	0.00000000 m	1×10^{-9}
Pico	0.00000000000 m	1×10^{-12}
Femto	0.000000000000 m	1×10^{-14}

(*Reference: Google—Explore the Nano World*)

Meaning of word *nano* as given in dictionary is exceedingly small, dwarf, etc. The ISO definition of nanoparticle (ISO-TS 27687) is of a particle between 1 and 100 nm (*Reference: BPE position statement on use of nanomaterials*).

Material particles may have various shapes such as square prism, prism, cylinder, plate, etc. If any one of the dimensions of particle reaches nano range (between 1 and 100 nm) then that material to which this particle belongs is termed as *nanomaterial*.

Table 1.4 is provided to give an idea of particle size of few materials.

Table 1.4: Particle size

Material	*Particle size (nm)*
Carbo black	10–80
Titanium dioxide	20–1000
Zinc oxide	20–1000
Silica	< 100
Iron oxide	50–1000
Chromium oxide pigment	300–400
Ultramarine pigment	300–800

(Reference: British Plastics Federation, literature on internet)

In nanoparticle, ratio of surface to volume is vastly increased. Properties of a normal material and with particle size more than 100 nm remain the same. If particle size of material reaches nano range then its properties suddenly change. For instance:

- Opaque substances become transparent (copper)
- Inert materials attain catalytic properties (platinum)
- Stable materials turn combustible (aluminum)
- Solid turns into liquid at room temperature (gold)
- Insulator becomes conductor (silicon)

Nanomaterials may be classified as under:

- Fullerenes
- Inorganic nanoparticles

Common method to produce fullerenes is to send a large current between two nearby graphite electrodes in an inert atmosphere. The resulting carbon plasma arc between the electrodes cools into sooty residue from which many fullerenes can be isolated.

The sol-gel process is a wet-chemical technique commonly used to synthesize a wide variety of nanomaterials.

(Reference, article : en.wikipedia.org/wiki/nano_materials)

Note: In above referred article, ***images*** *of various nano dimensional materials may also be seen.*

TEST

1. Atom of a chemical element is made of what parts? Describe with the help of a simple sketch.
2. Nucleus of an atom contains what? Describe with the help of a simple sketch.
3. What is bond of two chemical elements? Describe with the help of sketch.
4. Draw a figure to show bond of hydrogen and oxygen.
5. Explain with the help of figure monomer and polymer of carbon and hydrogen.
6. What is the difference in the nomenclature, polymer and plastics?
7. List out important properties of plastics and indicate units of measurement.
8. What is significance of density and melt flow index of plastics?
9. How many types of plastics are generally used in industry?
10. What is the value of prefix *nano,* if indicated in numbers? Briefly explain what is nanoplastics.

ASSIGNMENT

Note: Tick mark in block for correct option and cross remaining

1. Wax belong to *family* of:
   (a) Natural polymer .... ☐
   (b) Synthetic polymer .... ☐
   (c) Or both .... ☐
2. Compare polystyrene and polyethylene:
   (a) Polyethylene has a higher density .... ☐
   (b) Polyethylene has a lower density .... ☐
   (c) Polystyrene has a lower density .... ☐
3. Octet principle of bonding is:
   (a) Inner path electrons contribute in bonding of atoms .... ☐
   (b) Outermost path electrons contribute in bonding of atoms .... ☐
   (c) Eight electrons are necessary to establish a stable bond of atoms .... ☐

4. Thermosetting plastics can
   (a) Be re-soften by applying heat .... ☐
   (b) Not be re-processed .... ☐
   (c) Be softer than thermoplastics .... ☐
5. Wear resistance of plastics materials is its property:
   (a) To exhibit more resistance in sliding .... ☐
   (b) To exhibit less wear between two sliding surfaces .... ☐
   (c) To exhibit more wear resistance between two sliding surfaces .... ☐
6. Which of the following processing temperatures for ABS is more favourable?
   (a) 180ºC .... ☐
   (b) 310ºC .... ☐
   (c) 210ºC .... ☐
7. Which of the following plastics has highest tensile strength at break?
   (a) High density polyethylene .... ☐
   (b) Low density polyethylene .... ☐
   (c) Acrylonitryl butadine styrene (ABS) .... ☐
8. Which of the following plastics has good light transmittance?
   (a) Low density polyethylene .... ☐
   (b) Polyethylene terephthalate (PET) .... ☐
   (c) Polymethyl methacrylate (PMMA) .... ☐
9. Which of the following plastics is said to be engineering plastics?
   (a) HDPE .... ☐
   (b) ABS .... ☐
   (c) Polyamide .... ☐
10. Which of the following plastics floats in water?
   (a) Polycarbonate .... ☐
   (b) Polypropylene .... ☐
   (c) High density polyethylene .... ☐

2 Plastics Processing Techniques

INJECTION MOLDING

Machines

Most of the machine designs and systems were developed during last century. Pace of fast development was triggered during and after Second World War, when economic considerations were a priority.

Various injection molding machine manufacturers in many countries developed machines to cater to the needs of molders who produced a variety of items and technical components. Injection molded parts and components like gears, computer cabinets, keyboard parts, automobile components are produced by a combination of machines and molds. Such components are produced from various types and grades of thermoplastics which have significant influence on the design of molds and selection of injection molding machines.

In this chapter, designs of various types of injection molding machines are discussed, giving reasons due to which machines of various designs are made. As already pointed out, shapes, dimensions and types of polymers and weight of plastics parts vary very much from few grams to many kilograms. Thermoplastics like general purpose polystyrene, polyethylene, PVC, etc. to ABS, polycarbonate and nylon are generally used. Figure 2.1 gives an idea of types of plastics and other parameters like dimensions and weight, mold dimensions and weight associated with some listed products.

Some of the parameters which are guidelines for machine designers are listed below.

- Type of plastics
- Shot weight
- Dimensions of molds

- Weight of molds
- Type of machine operation
- Special purpose designs
- Auxiliary facilities
- Accuracy and reliability
- Competitive cost

On the basis of above guidelines, plastics injection molding machines are made. A number of types of machines are made. It is very essential to point out that machine manufacturers make only standardized machines. Mold designers take into account most suitable standardized machine while designing a mold.

Machines may be categorized as shown in Fig. 2.1.

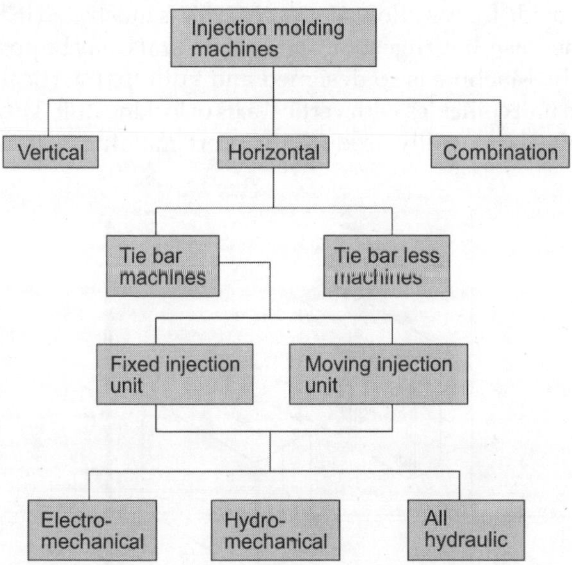

Fig. 2.1: Machine parameters for components

Vertical Injection Molding Machines

Plastic molding machines construction may be considered as combination of units which are as follows.

- Injection unit
- Locking unit

- Power/Drive unit
- Feed unit
- Control unit

These units are almost in all types of machines with the exception of tie bar less machine. Locking unit is there but without tie bars.

Vertical injection molding machines are normally used for small size molds. As the size of mold depends on size, weight and number of cavities, small components like bushes, washers, clips, caps, clock gears, mobile, electrical, electronic components, etc. are molded.

In a vertical molding machine, there can be option for keeping injection unit in vertical position and locking unit in horizontal position. In this case, injection of plastics is done through a vertical sprue made in one half or both halves of mold. Generally, full round sprue is made for easy flow of plastics. This is illustrated in Fig. 2.2. In another case, both injection and locking units may be positioned vertically. Machine is so designed and built that vertical axis of injection unit coincides with vertical axis of locking unit. This setting of machine is generally used where insert molding is done.

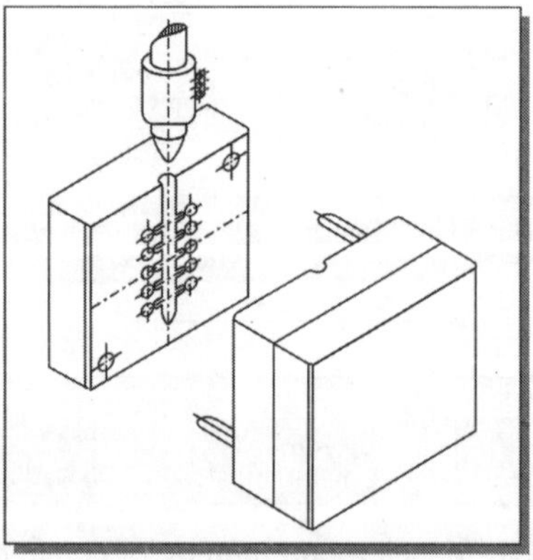

Fig. 2.2: Vertical injection

In this case meeting faces of mold halves are horizontal. Lower half of the mold has cavities or cores to receive insert which remains in portion due to gravity. Such is not the situation where cavity plates are in vertical position and hence there is always likelihood of inserts falling down out of core or cavity. Molds for this setting of machine have centrally located sprue through cavity plate. This typical arrangement is shown in Fig. 2.3.

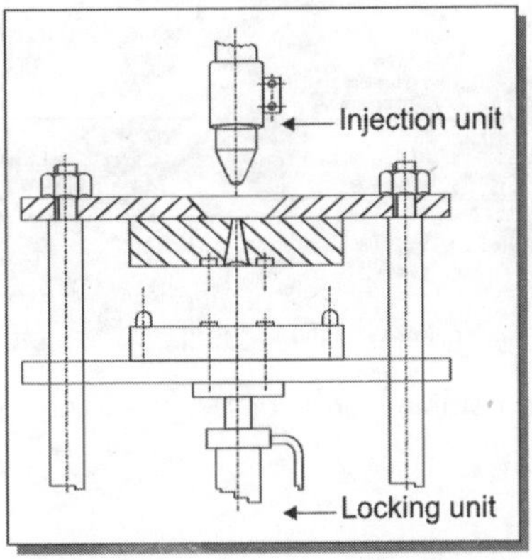

Fig. 2.3: Vertical mold opening

There are few convertible machine manufacturers who provide a third option of conversion in which both injection unit and locking unit are in horizontal position. This means that machine due to versatility in setting of injection and locking units, are light weight. Following websites may be visited on internet to have good idea of such machine.

www.plas-video.com/IMM/vertical_injection_molding_machine.htm

www.injection-molding-resource.org

www. Google.com   On address bar write injection molding machine convertible-google patents then Enter www.Google.com

On address bar write injection molding machine> then select images for injection molding machine-report images

Arburg Germany is one of the companies famous for making convertible machine (Photos 2.1 and 2.2).

Design and construction of injection machine as discussed above may also be mentioned under classification *combination* (*see* diagram) provided; it is possible to change position of injection unit

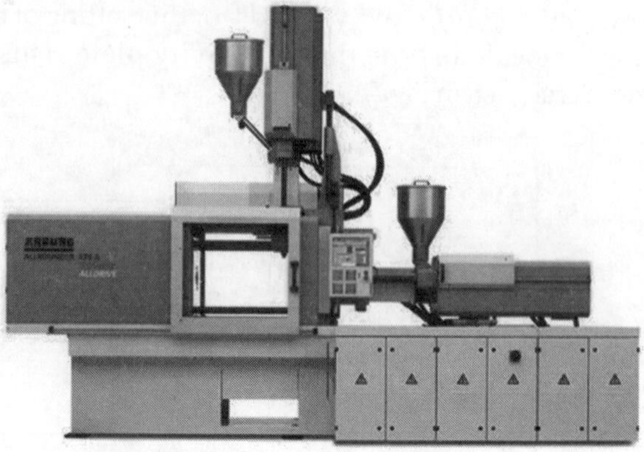

Photo 2.1: Arburg combination injection moulding machine *(Courtesy: Arburg Germany)*

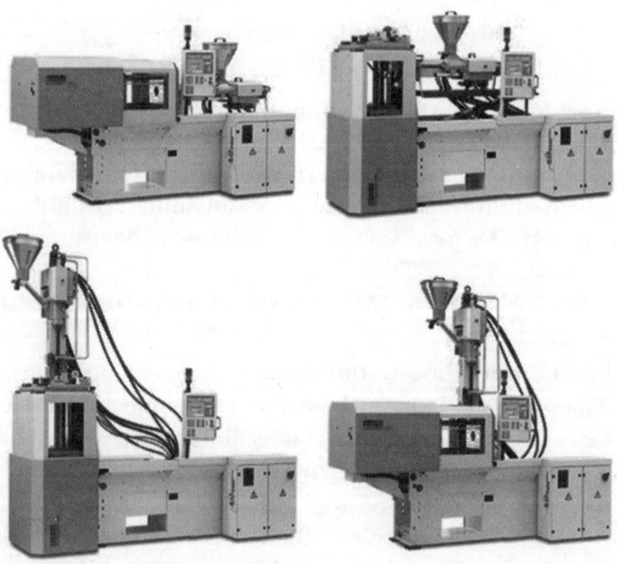

Photo 2.2: Four positions of Arburg injection molding machine *(Courtesy: Arburg Germany)*

from vertical to horizontal and vice versa while locking unit is in horizontal position.

Horizontal Injection Molding Machines (Photo 2.3)

Any injection molding machine may be classified as *horizontal,* if locking and injection units are permanently in horizontal position. These machines are designed and made with specification typically ranging from 50 to 2000 tons locking force and shot weight capacities ranging from few grams to few hundred grams. Dimensions and capacities of machine vary significantly with variation in shot weight, component design, mold functioning, size and dimensions. Table 2.1 provides an approximate idea as to what size, locking force and shot weight of machine would be suitable for loading molds of given parameters.

Photo 2.3: Windsor injection molding machine *(Courtesy: Windsor India)*

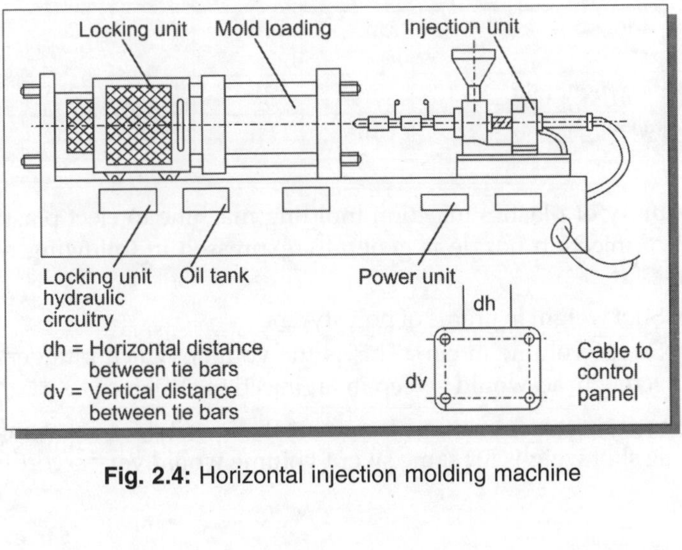

Fig. 2.4: Horizontal injection molding machine

Table 2.1: Machine parameters for components

S. no.	*Description*	*Unit*	*Components*		
			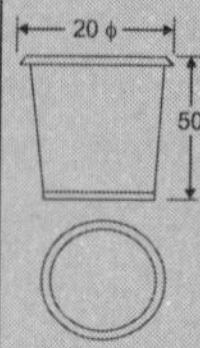	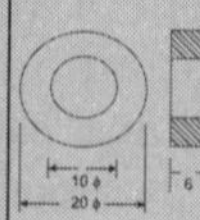	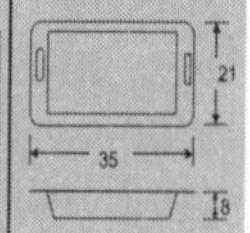
1.	Weight	gm	450	65	110
2.	No of cavities		1	2	1
3.	Shot weight	gm	460	140	120
4.	Projected area	sq cm	314	235	735
5.	Injection pressure	kg/cm^2	214	190	214
6.	Theoretical locking force	ton	265	213	620
7.	Mold length	cm	40	40	65
8.	Width	cm	40	40	40
9.	Height	cm	100	40	40
10.	Machine locking	ton	450	450	1200
11.	Shot weight	gm	1456	896	896
12.	Injection pressure	kg/cm^2	840	800	840
13.	Distance between tie bars	cm	70 × 70	70 × 70	119 × 119
14.	Mold opening stroke	cm	150	51	22
15.	Daylight min	cm	90	31	265
16.	Daylight max	cm	140	82	287

Ability of plastics injection molding machine to eject plastics out of injection nozzle is generally expressed in following two ways:

- Shot weight in grams of polystyrene.
- Swept volume in cm^3. This is the volume which plunger or screw head would sweep in an injection stroke.

Specifying swept volume is more realistic as it does not change. While shot weighs for same swept volume would vary according to density of plastics being processed. Shot weight would then be swept volume multiplied by density of plastics (Fig. 2.5).

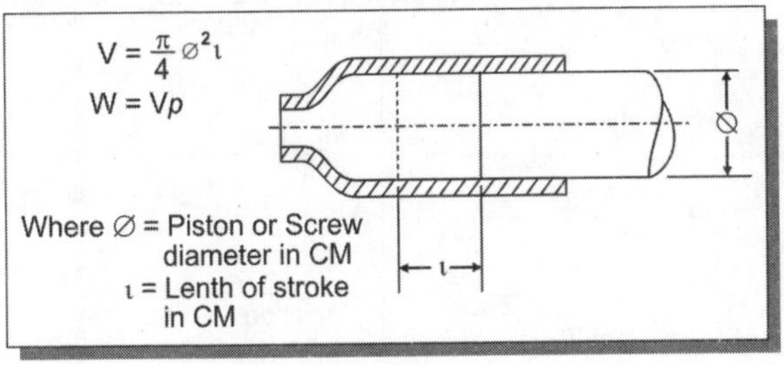

Fig. 2.5: Injection volume

Combination Machines

Reference of combination machine is already made while vertical plastic injection molding machines are discussed. Now here are some more details. Figure 2.6 shows general arrangement of injection unit feed (hopper) and locking unit. Injection unit can be swung from horizontal to vertical position. This can be done by mechanical arrangement provided in this machine. Once in vertical position, it is locked rigidly to remain in this position. At the same time, the direction of hopper is also changed to remain upright (Fig. 2.7). Other option may be that both injection units and locking units are in vertical position, *see* Fig. 2.4 for three alternative positions of injection and locking units.

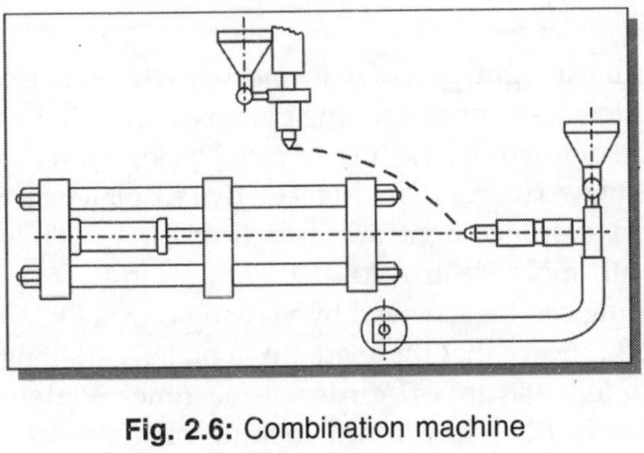

Fig. 2.6: Combination machine

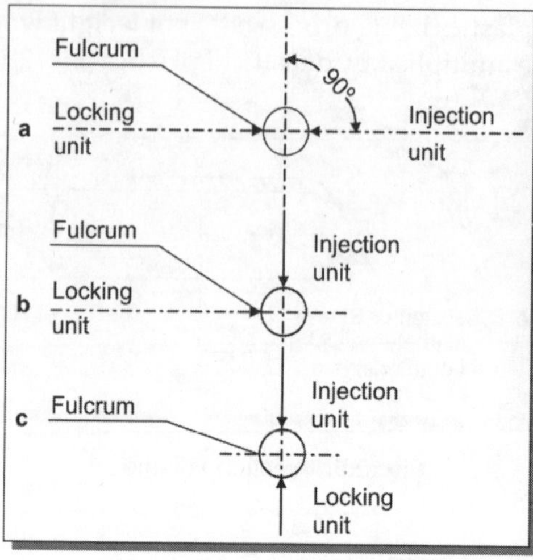

Fig. 2.7: Combination diagram

Tie Bar Machine

Following aspects of plastics injection molding machine tie bars are described as below.

- Purpose of tie bar
- Tie bar steel
- Design
- Heat treatment
- Surface finish
- Accuracies

Purpose of Tie Bar

Tie bar became integral part of machine, when fitted to fixed platens of the machine. It supports moving platens of machine which carry moving side of mold. At the time of mold locking force generation, when faces of two halres of mold meet, two fix platens would come under great displacement and bending forces. These platens together with mold destroyed, if there are no tie bars. A better understanding can be achieved by viewing (Fig. 2.8).

Figure 2.8a shows that there are three platens: Platens P_1 and P_2 are fixed to longer support girder of machines. Platen P_3 would move towards P_1, when mold locking jack works. It can be imagined that generated locking force would try to tilt platens P_1 and P_2 and get displaced resulting in destruction of platens, their supports and even mold may be damaged.

Figure 2.8b shows that there are four tie bars to keep fixed platens in place, when mold locking force is generated. Hence there would be no destruction of platens, their support and mold.

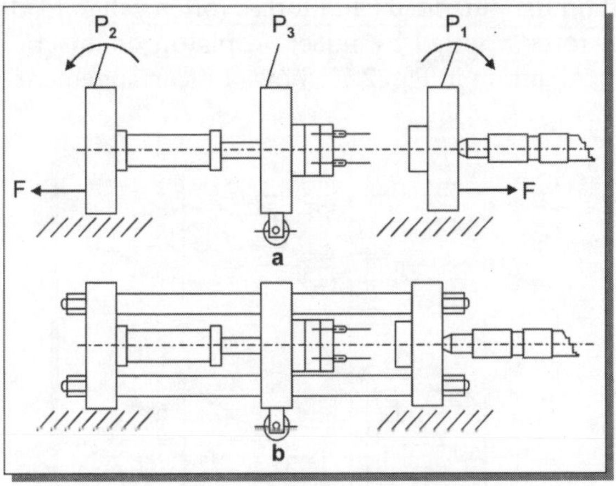

Fig. 2.8: With and without tie bars

Tie bars in an injection molding machine may be two or four. In earlier small capacity machines, two tie bars could be seen. But nowadays four tie bars are normally used. Tie bars may be of length to hold two fixed platens as shown in Fig. 2.8b. Tie bars may be longer even to hold fixed platens of injection unit in addition to holding other two fixed platens. This is shown in Fig. 2.9.

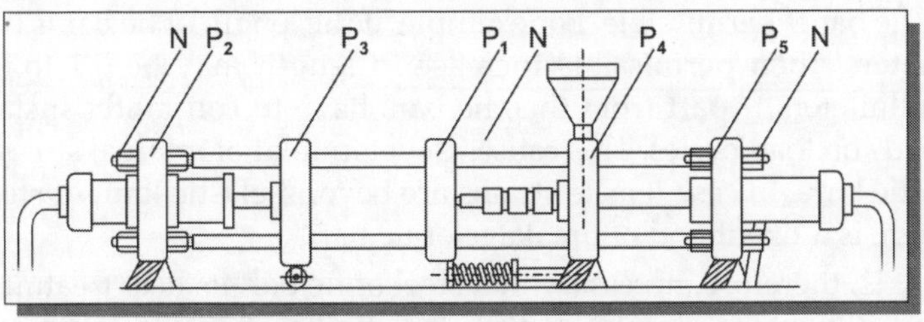

Fig. 2.9: Injection molding machine

These are platens (4) and (5). Platen (4) holds plasticizing cylinder and plastics granules feed systems. Platen (1) moves on tie bars for a short distance at the mold locking and rest against robust tie bar nuts N capable of sustaining mold locking force which is evenly shared by four tie bars via robust nuts.

In some machine designs, tie bars for injection unit are different to take load of plasticizing cylinder, granules intake system, plasticizing screw rotation system and hydraulic jack system. These tie bars sustain hanging loads and forces (often called loads) due to injection force applied by injection piston or injection screw. Schematic diagram in Fig. 2.10 shows this arrangement.

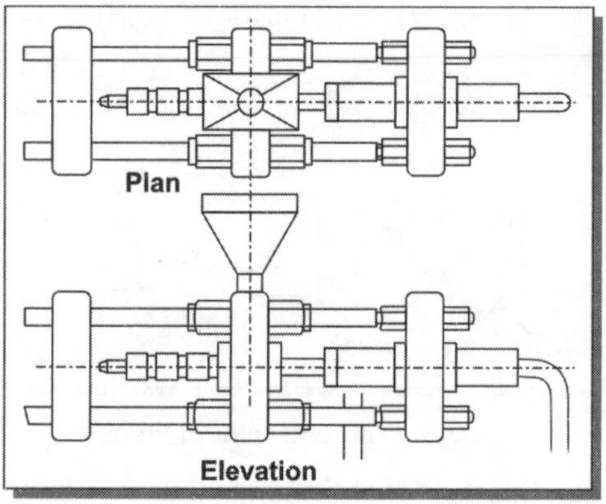

Fig. 2.10: Machine platens and tie bars

Tie Bar Steel

During locking stroke, locking forces tend to elongate tie bars, which should not take place. Only a very small increase in length of tie bar is permissible. For example, if the length of tie bar is two meters' then permissible increase in length may be 0.1 to 0.3 millimeters. Apart from this, tie bars have to constantly sustain load/no load cycles. This causes development of internal stresses in tie bars. In case tensile strains are beyond elastic limit of steel, there is a likelihood of breakage of tie bar.

It is, therefore, necessary that steel of tie bar, its heat treatment and dimension should be such that tie bar remains intact and serviceable for almost complete lifetime of machine.

Tie bars have to undergo tensile, shear, bending, shifting load factors and fatigue. Tie bar designers take into account all these factors with the objective of *fail safe* design. Exhaustive knowledge about composition, grain structure and mechanical properties of steel is necessary for designing a tie bar. It is further desirable that designer should have knowledge about past performance of tie bars on shop floor. This provides an insight to designers as to which type and grade of steel gave most satisfactory performance.

Designers often consult latest steel specification sheets provided by steel makers. A typical steel specification sheet is provided in the form of Table 2.2.

Table 2.2: Steel specification

Mechanical properties

S. no.	*Metal*	*Modulus of elasticity million psi*	*Ultimate tensile strength thousand psi*	*Yield strength in psi*	*Endurance limit in thousand psi*	*Hardness Brinell*
1.	Wrought plain carbon steel C 1045 hard and temper	29–30	118	88	–	277
2.	C 1095	29–30	180	118	–	375
3.	Ultra-high strength steel H_1	30	295–311	241–247	132	–

On studying the properties of steel given in Table 2.2, selection of properties of ultra-high strength steel H_1 may be done for calculating the diameter of tie bar. This calculation would take into account a safety factor of say, 3. This means cross-section of tie bar would be kept three times the calculated value. It may be considered enough to give a long-life to tie bars.

Figure 2.11 shows the threaded end of a typical tie bar. Length of threaded portion is kept so much that it is possible to tighten platen with the help of two nuts. Nut which has to take locking load is made extra long. Threads are generally square or trapezoidal. These threads are so designed that there is no chance of failure due to shearing of threads.

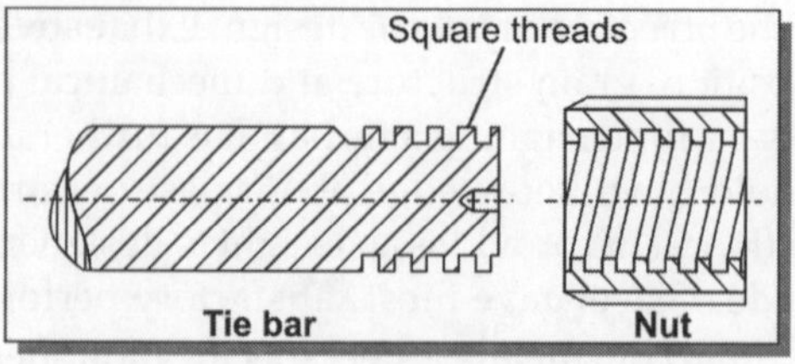

Fig. 2.11: Tie bar threads and nut

Tie Bars Less Injection Molding Machine

There are no tie bars for locking and injection units. Both units move in guide ways built on robust 'C' type cast steel body. These types of machine are rarely used for special reasons. Figure 2.12 shows schematic shape of machine.

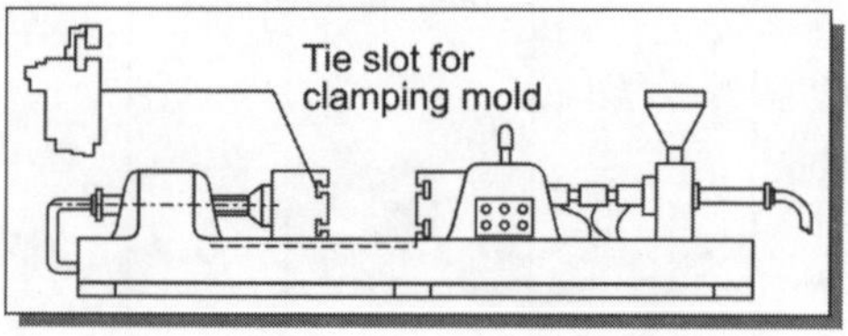

Fig. 2.12: Tie bar less machine

Fixed Injection Unit

Purpose of an injection unit is to plasticize plastics to desired level of plasticization and then to push plastics with adequate pressure and speed into the mold. This can be done, only when face of injection nozzle and mold feed head face are firmly in contact with sufficient force so that plasticized plastic does not leak out while injection is taking place.

At the same time, it is also desirable that contact between cylinder nozzle and mold sprue bushing should not remain permanently in contact with feed head to avoid cooling down of nozzle due to which formation of drawn plastics strand takes place while mold is opened with runner and sprue. For this reason, generally machines are designed and built in following two ways:

1. Fixed mold platen and moving injection unit
2. Moving mold platen and fixed injection unit

Moving Injection Unit

Option *moving mold platen* is almost out dated due to process limitation and difficulties in platen setting with mold already loaded on it.

Option *fixed mold platen and moving injection unit* is used in modern injection molding machines. Advantages of moving injection unit are as below:

- Nozzle to feed head contact pressure can be adjusted according to injection pressure and robustness of mold, specially feed head/sprue bushing.
- Working space between injection nozzle and back of platen can be increased for ease of doing work on nozzle or replacing band heaters.
- Setting of cycle timing and sequence.
- No disturbance to locking unit or locking tie bar in case of dismantling of injection unit for repair or preventive maintenance.

In some injection molding machines, injection unit moves in guide ways built on machine structure. Hence, there are no tie bars for injection unit. General arrangement is shown in Fig. 2.13.

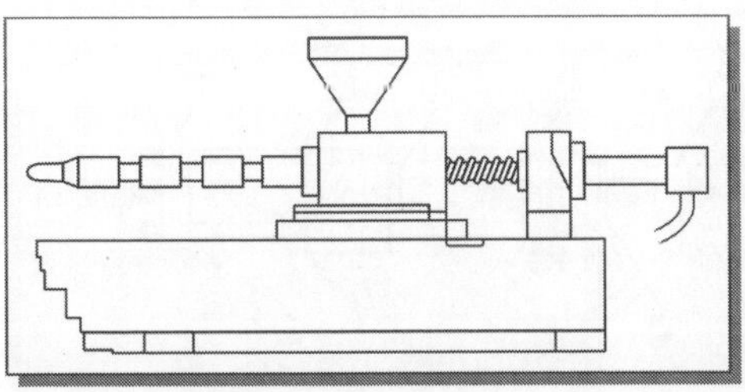

Fig. 2.13. Injection unit

Electro-mechanical Injection Molding Machines

In this type of machines, movement of locking unit and injection unit is directly achieved from electric motors through various gears, linkages and toggle joints. Sequence of operations, start/stop limits, timing and safety action are achieved by electrical circuitry, timers, limit switches, etc. Locking force is adjusted by mechanical robust nuts and screws. Injection pressure is adjusted by hydraulic plus gas pressure system.

Hydro-mechanical Injection Molding Machines

In these machines, various types of locking system are adopted, which are as below.

- Toggle joint is operated by hydraulic jack (Fig. 2.14).

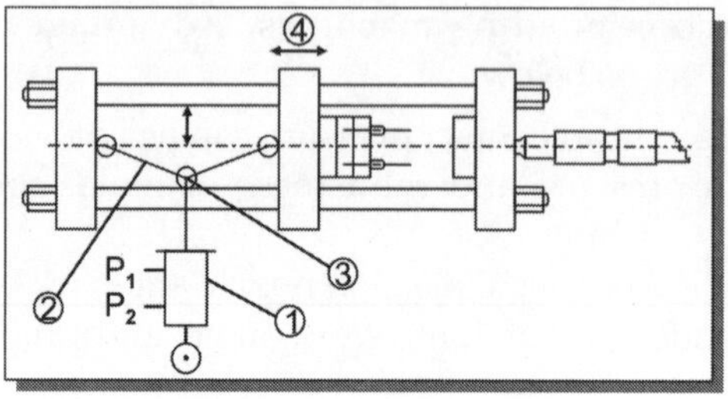

Fig. 2.14: Platen actuation

- Moving platen is big and contains hydraulic pressure built up system. This platen is moved forward and backward by means of robust threaded cylinder which is made to move axially by means of a rotating nut, driven by electric motor. When mold gets closed, locking pressure built up (Fig. 2.15).

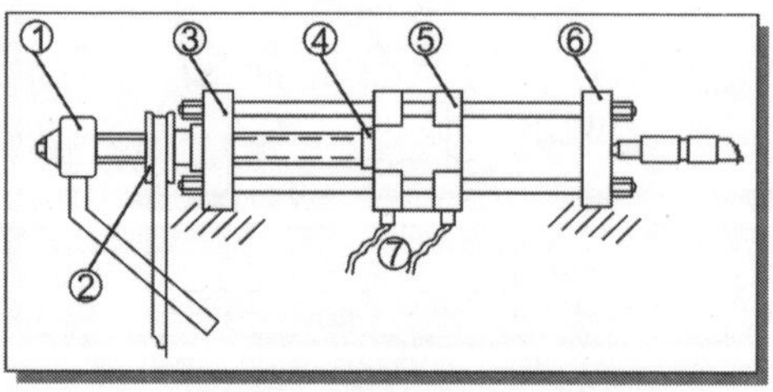

Fig. 2.15: Locking system

- Moving platen moves fast to get the mold closed. This is achieved by hydraulic cylinder built around tie bars and connected to moving platen. After platen has moved completely to lock the mold with low force, there is a gap between back of platen and high force application jack face. A gap filling steel block swing in the gap and full locking force get applied. Sequence of all these actions is controlled by machine operation setting. Schematic arrangement is shown in Fig. 2.16.

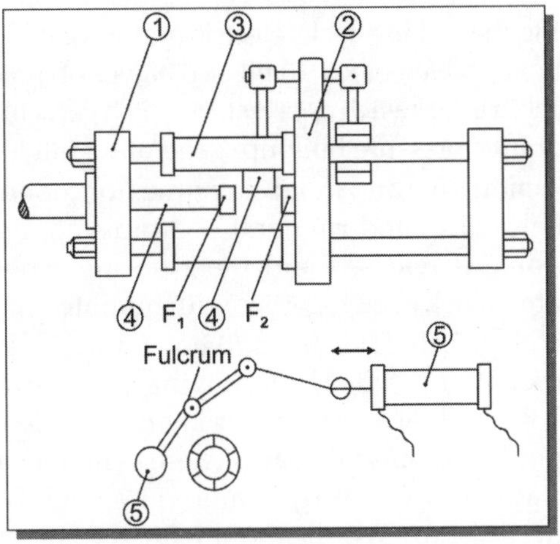

Fig. 2.16: Swinging spacer locking system

Referring to Fig. 2.14, (1) is a hydraulic jack having two ports, P_1 and P_2. When direction of flow of hydraulic oil is into port P_1 then piston rod moves downwards, bringing toggle joint (3) down.

If the flow of hydraulic oil is into P_2, then piston rod moves up, thus moving toggle joint (3) up wards so much that links (2) take exact horizontal position. Consequently, platen (4) horizontally moves backward and forward to open and close mold. It is to be noted that when links (2) are about to attain horizontal position, full mold locking force is generated.

Referring to Fig. 2.15, (1) is a big diameter hollow rod. It has square or trapezoidal strong threads throughout its length. This hollow rod is fixed to hydraulic piston rod (4). This piston rod is also big in diameter and is hollow. Full high pressure hydraulic jack system is housed inside moving platen block (5). Hollow of screw and jack rod allows to pass ejection system, that is also hydraulically operated. Whole mold locking cycle is as below.

- Rotating pulley (2) is axially connected to fixed platen (3). This pulley can rotate but cannot move axially. There are thread at the center from which threaded hollow rod (1) passes. Clockwise and anticlockwise rotations of pulley make screw to move forward and backward.

- On starting the locking cycle, electric motor rotates pulley (2). In turn, it moves platen block (5) till two halves of mold get closed. At this time, a limit switch or pressure switch get actuated causing high hydraulic pressure built up, generating full locking force.
- After injection time and mold stay time (for cooling of plastics inside the cavities and runners) is completed, high pressure locking force is released and then locking screw rotates to bring platen block back together with mounted halve of mold.

Referring to Fig. 2.16, (1) is fixed platen on which big diameter hydraulic jack (4) is fitted. (2) is moving platen which carries moving half of mold. Movement of platen is on tie bars. (3) are hydraulic cylinders around tie bars. These cylinders are attached to moving platen (2). When hydraulic oil enters the cylinders, platen (2) moves fast to close the mold with normal pressure. Due to shifting of platen, a gap is created between face F_1 of high pressure hydraulic jack piston and F_2 back face of platen (2). (5) is a swinging spacer capable of swinging in the gap.

- As soon as spacer reach completely in the gap and high locking force is applied on platen (2).
- Mold halves are now locked to receive injection of plastics.
- All the movements are sequenced and timed by micro-processor or PLC (programmable logic control) of machine.

All Hydraulic Machines

In *all hydraulic* machines, typically there are three hydraulic circuits:

1. Low pressure control circuit
2. Normal pressure circuit for rotary and linear motions
3. High pressure for mold locking and injection

Figure 2.17 gives an idea of electrohydraulic circuitry.

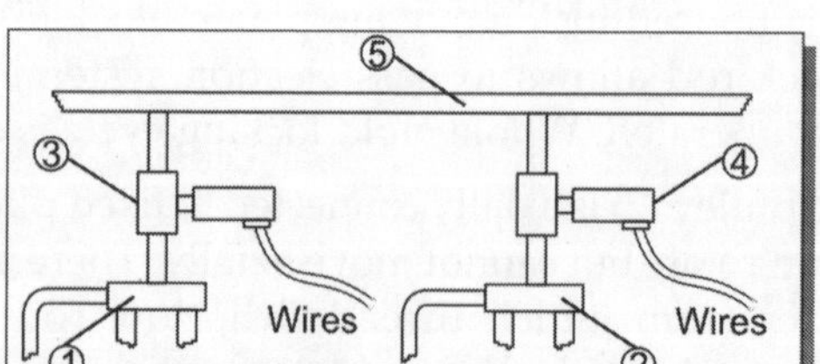

Fig. 2.17: Hydraulic oil flow

There is a low pressure hydraulic line (5). It is meant to operate main direction control valves (1) and (2). Flow of low pressure oil is controlled by solenoid valves (3) and (4). Hence action of main direction control valves depends on sequence and timing of energization of solenoid valves which are controlled by micro-processor of machine.

Figure 2.18 provides an schematic sketch of locking pressure intensifier. (1) is cylinder of main locking jack. (2) is an auxiliary pressure intensifier jack. Its piston rod (3) is small in diameter. Small length of rod passes through high pressure oil seal so that hydraulic oil does not leak from one cylinder to other. Let us assume that cylinder (1) is filled with hydraulic oil at normal locking pressure and mold is closed with low locking force.

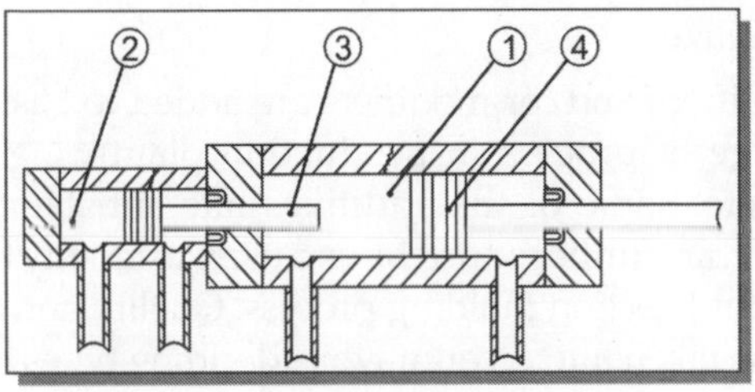

Fig. 2.18. Locking force intensifier

Electric and hydraulic circuits allow hydraulic oil to generate pressure in cylinder (2). This happens at appropriate moment of sequence and timing. Piston (3) tends to enter into cylinder (1) which is already full with hydraulic oil under normal pressure with mold in low force lock position.

Since the diameter of piston (3) is much less than diameter of piston (4), force exerted by piston (3) on piston (4) causes intensified pressure in cylinder (1). This generates very high pressure causing mold locking with full force.

Preparation of Plastic Granules and Pellets

Polymers producing companies normally produce plastics materials in the form of granules by cutting plastics extruded strands. Pellets are oval-shaped smooth grains and its production is in the last stage of polymer production process.

Generally, plastics granules or pellets are to be prepared for feeding into injection molding machine hopper. Following are few additives or modifiers which are generally mixed.

- Anti-blocking agents
- Antimicrobials
- Antioxidants
- Antistatic
- Carbon blacks
- Colorants
- Flame retardants
- Heat stabilizers
- Light stabilizers
- Lubricants
- Plasticizers
- UV stabilizers

Many additives and/or modifiers are added to basic polymer at suitable stage of process in production plants, e.g. plasticizer, lubricants, etc. Some of the additives like antistatics, colorants, lubricants, etc. are further added by end users to modify the properties of granules for injection molding process. Quality norms of molded items are also taken into account while deciding types and grades of additives. For example, elastomer is added to polystyrene to make it impact-resistant which is called high impact polystyrene (HIPS).

Plastics like polyamide, HIPS, plycarbonate absorb moisture from atmosphere. Polyamides are hygroscopic polymers and need vacuum driers for efficient drying. ABS and many other plastics retain moisture on granule surfaces. It can be dried in conventional ovens. This may be the case with some other plastics as well. Simple heating in ovens would drive out moisture from the surface of granules. Once granules are dried, then master batches and additives are evenly mixed before feeding to machine hopper.

Nowadays, majority of machines are fitted with screw plunger plasticizing and injection system. In these machines, natural color granules are directly fed to machine hopper where metered quantity of color pigments, master batches or additives is kept on feeding to coming in granules. Rotary and pushing actions of screw plunger mix plastics granules and additives together to make them a homogeneous mix before it comes in front of screw head, ready for pushing softened plastics out to mold.

Feeding of Plastic Granules to Injection Molding Machine Hopper

There are two ways of feeding plastics granules into plastics injection molding machine hopper. In small capacity conversational machine, feeding of granules is manually done. In this case, operator of machine can do this job by taking out granules from bag in a small container and then to drop granules into the hopper. One filling in hopper may be about 10 kg which may last for about 100 minutes, if shot weight is 20 grams. In a molding cycle of 5 shots per minute, operator has also to care for removal of any contaminant incidentally present with plastic granules. In case of manually operated machine, operator may not have time for loading of plastics materials in hopper. Another worker would have to be assigned with plastics granules loading job. This way of feeding granules is outdated and rarely adopted. In modern molding shops, feeding of plastics granules is carried out with automatic hopper loading system. Figure 2.19 shows a typical automatic hopper loading system for only one machine.

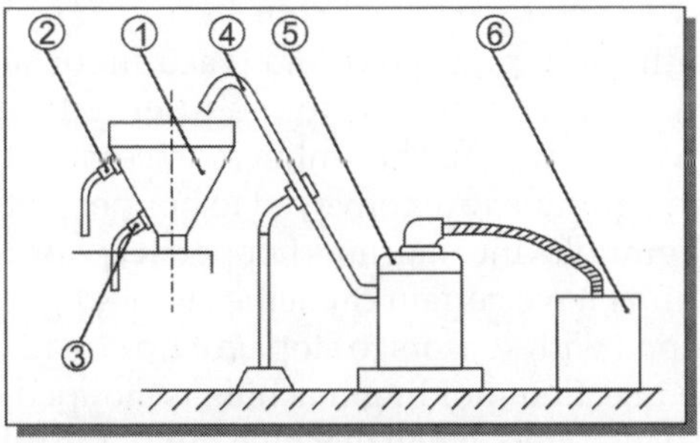

Fig. 2.19: Hopper loader

(1) is the hopper of a plastic processing machine, an extruder or an injection molding machine. Outer surface of hopper is fitted with electronic sensors. There are small openings in the wall of hopper. These openings are covered with glass plates. Over these glass plates, sensors are fitted. Upper sensor (2) triggers switching off of vacuum loading unit (5). As soon as level reaches sensor (3), it triggers restart of vacuum loader. There is also an audio alarm system provided in vacuum loader which gives an alarm, if

granules bag become empty. It is then replaced by another bag filled with plastics granules.

In a large plastic processing plant, there may be a number of machines which are to be fed with plastics granules almost round the clock. Feeding of granules to all the machines may be constant or intermittent, depending of running hours of machines.

Figure 2.20 shows schematic arrangement of transport system for transporting granules from silos to machine hopper.

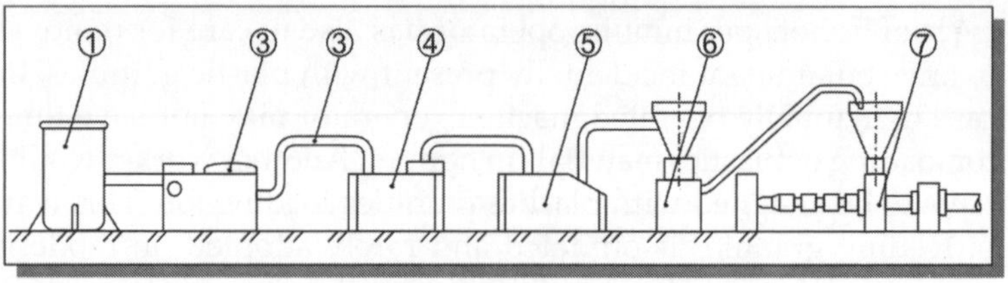

Fig. 2.20: Granules loading system

(1) is plastics granules silos where bulk quantities are filled from rail or road special carrier. From here, granules are then transported through pipes to driers (vacuum or non-vacuum). Granules are then transported to modifier/additives unit (4) through conveyor pipe (3). Then inline, comes master batch mixer (5). From here, granules are conveyed to hopper loader (6) which in turn loads granules into machine hopper. It is worth mentioning that all the units have certain capacities to hold granules. Every unit is equipped with sensors to stop unit up stream, if gets filled up. Suppose injection molding machine is stopped then hopper loader would also stop, when machine hopper is full. In turn, all units would stop and remain stopped till granules level comes down.

Plasticization of Granules

Granules of plastics cannot be injected into molds sprue, runners, gates and cavities as granules are normally hard (hardness depends on type and grade of plastics) and is not one homogeneous mass. For injecting thermoplastics into mold cavities, it is necessary that granules are softened and made a homogeneous mix by applying heat and churning motion. This is done inside plasticizing cum injection unit of injection molding machine. Figure 2.21 shows typical stages of granules being softened and injected into mold cavities.

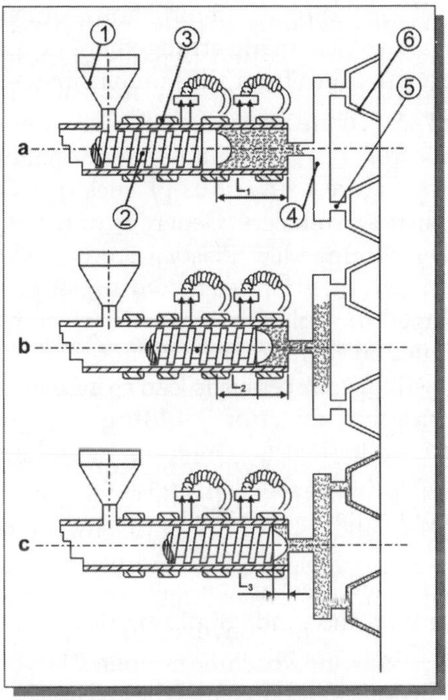

Fig. 2.21: Progressive cavity filling

(1) is hopper. (2) is screw plunger. First of all it rotates, causing granules to advance and gradually being heated with the heat supplied by heaters (3). While screw is rotating, plasticized plastic starts getting collected into space L_1. Which increases to L_2 while plasticized plastics moves into space. Screw axially moves back to make room L_2 for plastics coming in front of screw and get collected in space L_2. This space is preset by machine control system. Complete filling of space L_2 is normally done while cylinder nozzle is already in contact of mold sprue bushing. After returning of injection units, no leakage of plastics from nozzle takes place as plasticized plastic infront of screw plunger head is not

under pressure to such an extent that it causes leakage. Figure 2.21c shows that with forward movement of screw plunger, plastics melt advance into mold runner. Due to continued application of injection force, plastics melt further advance to fill cavities as shows in Fig. 2.21c. This completes the injection cycle. It can be seen in Fig. 2.21 that heating cylinder is fitted with band heaters. Heat from these band heaters soften plastics granules. Heating is gradually increased, held or reduced by mean of controlled heating zones. For each zone, there are temperature sensors which activate temperature controller to maintain temperature at preset values. Processing temperature for various plastics types and grades is specified by plastics producers. Generally, a range of processing temperature is recommended. Plastics producers determine this processing temperature range in two or more ways. One is complete designed formulation of plastics to be produced. This includes production technique to be adopted. Other method adopted by plastics producers is to carry out exhaustive tests by actually carrying out injection molding of plastics which is intended to be used for injection molding by end users. Similarly, processing temperature is determined for other processes as well, like extrusion of lay flat tubing, sheets, strands or blow molding, etc.

Plastics injection molders at one stage or the other set machine for a particular type and grade of plastics for the first time while setting plasticizing temperature of heating cylinder may take a start with recommended processing temperature. Then by trial and error, processing temperature for best molding results may be obtained. This processing temperature may be recorded and also fed to PLC (programmable logic control) for future use of this particular type and grade of plastics and mold.

Injection of Plastics into Mold

For proper understanding of injection of plastic into mold, it is necessary to know sufficient details about plastics flow path and entry into the cavity. Following are stages for which discussions would be done.

- Nozzle to feed head contact
- Mold with runners
- Mold without runners

- Single and multicavities mold
- Hot runners molds
- Type of gates
- Position of gates into cavities

Figure 2.22 shows a plasticizing cylinder nozzle (1) and mold feed head (2). Mold setting is done in such a way that axis of sprue (hole in feed head) (2) coincides with the axis of nozzle (1). Face of feed head (2) has a concave surface with a radius R_1. This concave surface is exactly concentric with sprue (hole in feed head). Similarly face of nozzle (1) has a convex surface with a radius of R_2. Center of convex surface is exactly concentric with nozzle hole.

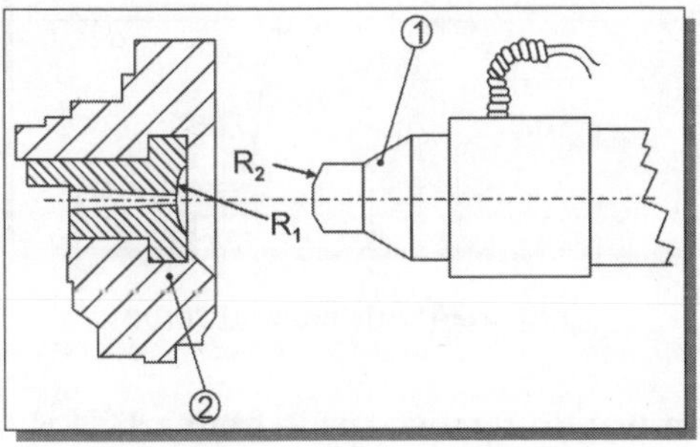

Fig. 2.22: Nozzle and feed head

Figure 2.23 shows two possibilities of matching of feed head and nozzle surface. Matching condition of surface shown in 'a' occurs, when R_1 is less than R_2. Condition shown in 'b' occurs, when R_1 is more than R_2. Condition shown in 'a' is highly undesirable as a thin layer of plastics would form which may not

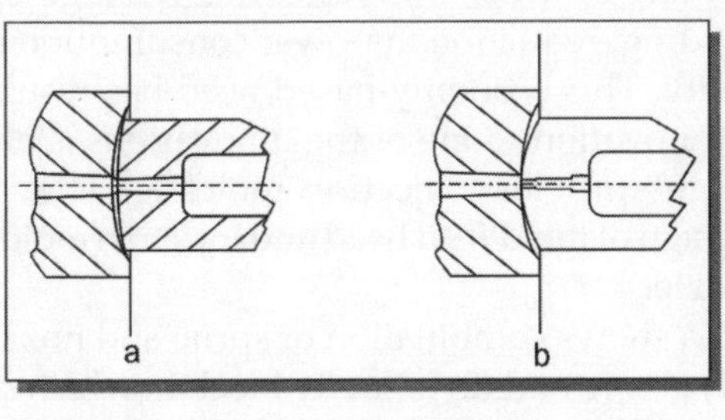

Fig. 2.23: Nozzle and feed head matching

allow removal of sprue, when mold opens and ejection of sprue and runners is to take place. Condition shown in 'b' is desirable. A third condition may be that both radii are exactly equal. This is rather difficult, when matching a feed head. That is why condition 'b' is said to be desirable.

Combination of feed head and nozzle described above is generally used for big articles like household bucket, tub, etc. Such items come out of mold with sprue attached to it as shown in Fig. 2.24.

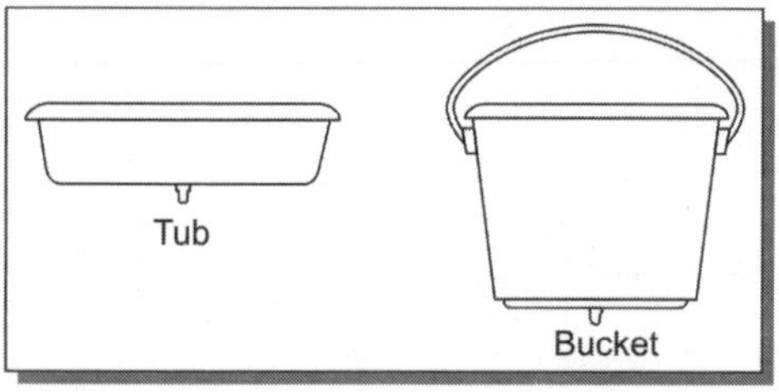

Fig. 2.24: Articles with sprue

Old fashion way to remove sprue from article is that operator immediately removes it manually or by a jig, when sprue is already sufficiently warm. In items like tubs and bucket, a large sprue of about 8 mm diameter is necessary because of the size of items. Plastics has to travel from center to brim. Therefore, fast filling is necessary to avoid cooling down of moving plastics inside space between cavity and core. Cooling down means that temperature goes down so much that further movement of plastics is not possible.

Since last three decades, there has been a surge for reducing cost of products by way of eliminating wastage of plastics in injection molding, reduction in power consumption and increase in productivity. This economy-based need has resulted in many technology innovations. One of the innovations is elimination of sprue through sprueless injection molding. This necessitated improved design of mold feed head and injection molding machine injection nozzle.

Figure 2.25a shows combination of sprue and nozzle design. In Fig. 2.25a, (1) is core and (2) is cavity block in which conical space (3) is provided. Closed end of cone is very close to cavity surface, say 0.5 to 1.5 mm. Through this thickness of steel, there is a small orifice or gate (5). Face of injection nozzle (4) remains permanently in touch with back face of cavity block. Conical space (3) is made in a specially designed steel insert of appropriate design to have insulation, heater, temperature sensor and cooling channels. Plastics in conical space in injection remain soft state while interrupted molding cycle keeps on repeating. Hence continuous production goes on. When production is stopped after completing the quantity, then plastics is allowed to cool down and then conical-shaped plastics form is extracted after withdrawing the nozzle.

Design of nozzle and gating shown in Fig. 2.25b is an improvement over 2.25a. It can be seen that instead of a full conical space, nozzle reaches inside conical space with all round uniform gap where plastics remain collected which form a thermal barrier between nozzle and conical space.

Figure 2.25c shows an arrangement where face of nozzle exactly flashes with cavity surface and there is no sprue. When mold opens, article (say a bucket) gets detached from nozzle orifice. Hence there is no sprue to cut.

It is worth mentioning that precision control of temperature at the tip of injection nozzle and mold surface around nozzle is very important. Therefore, very sensitive and precise temperature sensor and controller is a must.

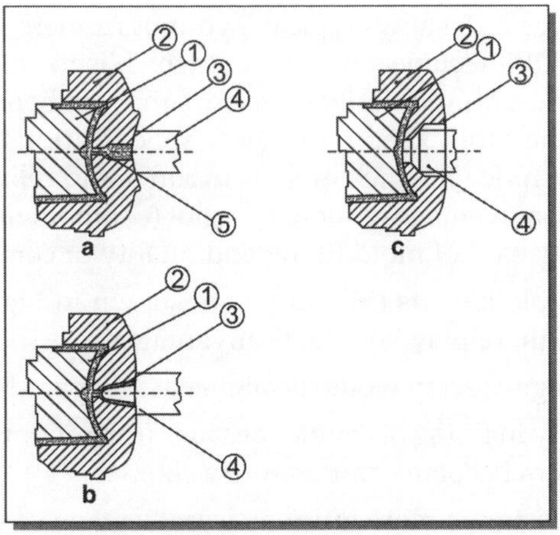

Fig. 2.25a: Types of nozzle

There are thermoplastics which become syrupy, when heated to a processing temperature of injection molding. Conversational nozzles are not suitable for these types of plastics as it will drool from nozzle orifice. Polyamide is an example of such a thermoplastic. Therefore, a special type of nozzle would have to be used where nozzle orifice remains closed from inside and opens only when injection pressure gets built up inside the nozzle.

Figure 2.25b shows typical design of such a nozzle.

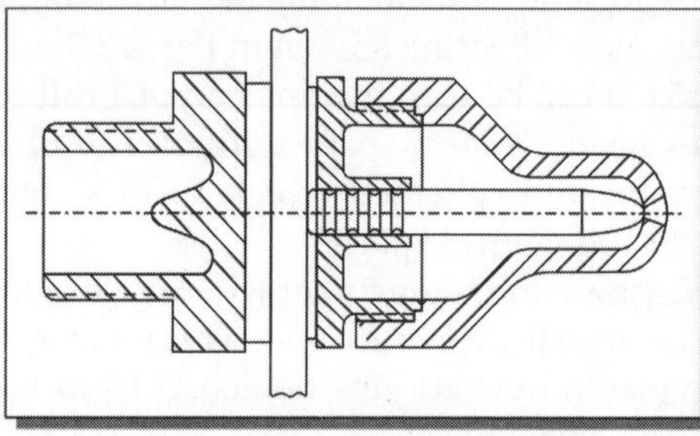

Fig. 2.25b: Self-closing nozzle

Runners are the pathways for plastics to travel up to gates of cavities and then to enter into the cavities. Runners are designed according to desired layout of cavities and temperature considerations. Figure 2.26 shows typical layout of runners. Figure 2.26a shows an HDPE technical component which is to be produced in bulk quantities. Mould designer decides the number of cavities to be kept in the mold keeping in view specifications of injection molding machine on which mold is to run, mold reliability, mold making cost, percentage of scrap in a shot (component plus sprues and runners), ease of mold filling and quality of components.

Two possible layouts (b) and (c) are shown in Fig. 2.26. In (b) type layout, there may be two disadvantages

1. Percentage of scrap would be higher as compared to layout (C)
2. Difficulty in filling of cavities becaues front of flowing plastics in runners become comparatively cool.

(C) type design is more efficient as percentage of scrap would be less and there are *cold wells* to *arrest* comparatively cool plastics and thus more hot-plastics would enter the cavities.

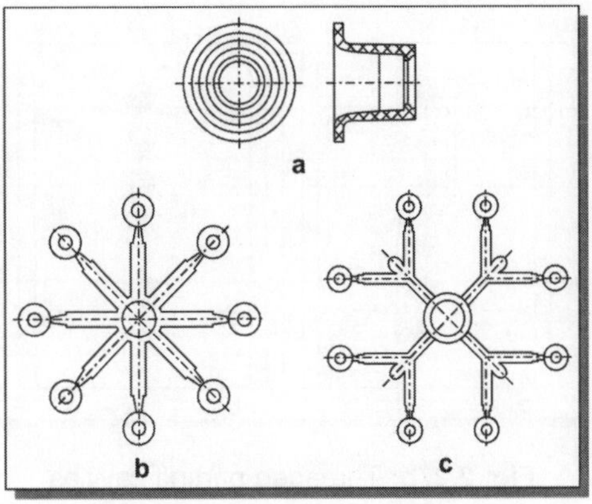

Fig. 2.26: Cavities layout

Figure 2.27a shows a technical component which is to be produced in bulk quantities. This component has a threaded portion. Cylindrical portion of components would be in injection side of mold cavities plates and threaded portion would be formed in the moving sides of mold. Both the portions of cavity are precisely concentric.

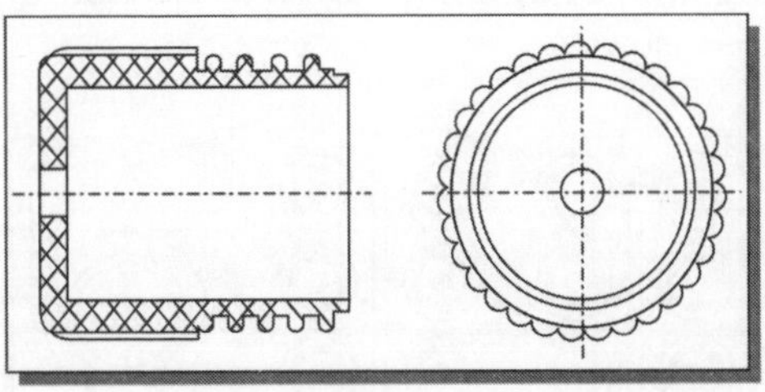

Fig. 2.27a: Threaded component

It is evident from Fig. 2.27b that cavities do not have a balanced location. *Not balanced* means that plastics do not reach cavity gates simultaneously. In this particular case, cavities would have to be in line as cavities plates (2) are to open during mold opening stroke to release thread portion of components to be ejected out of cores.

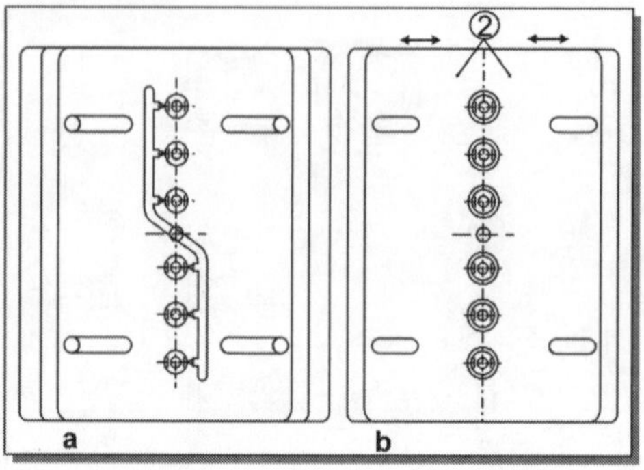

Fig. 2.27b: Threaded parting cavities

Figure 2.28b shows an *H style* runners layout for a component shown in Fig. 2.28a. Although runner is of unbalanced configuration but it could not be helped to make a moderate size of mold for 24 cavities. Mould designer decides to adopt a particular runner layout after taking various variable considerations into account.

Removal of components from runners is always a problem for molding shops as it involves labor cost and variation in the quality of degating. Consequently, method of *self-degating* during mold opening and/or ejection stroke of ejection in the mold is

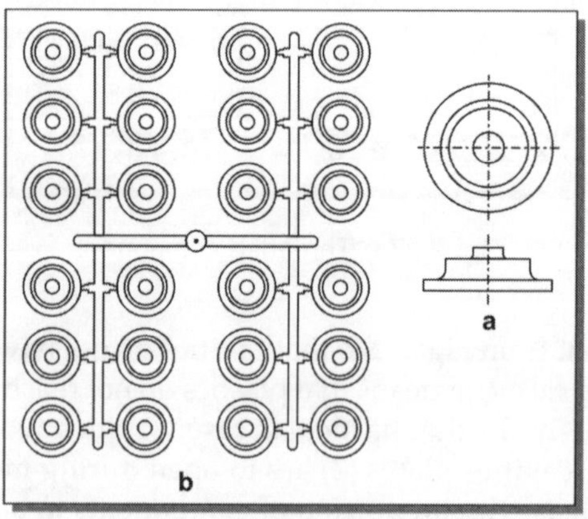

Fig. 2.28: H-type cavity layout

devised. Once again it led designer to re-allocate runner in consultation with product designer. Let us reconsider location of gate on component and runner layout in Fig. 2.28. It is more permissible to have a pin point gate at centre point on the top of components. This necessitated design of a 3-plate mold where runner is located between fixed injection side mold plate and floating cavity plates. Schematic view of this arrangement is shown in Fig. 2.29.

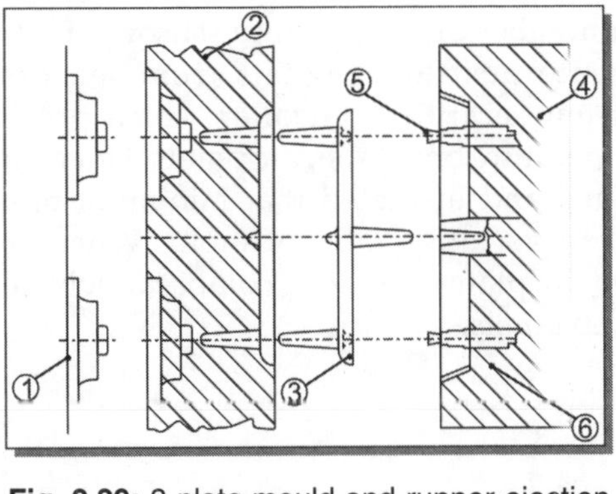

Fig. 2.29: 3-plate mould and runner ejection

(1) is stripper plate of moving half of mold which carry cores. (2) is floating cavity plate. (3) is runner (scrap). (4) is fixed plate of injection side of mold. Retaining pin (5) keeps the runner held in position at the time of mold opening. As soon as plate (2) moves for opening, pin point gate gets snapped from components. This happens because runner is held back by pin 5. After almost complete opening of mold, stripper plates (6) moves forwards to release runner from pin 5. In this way, runner (3) is either fells down by its own weight or a robot is employed to remove runner from mold.

Basic objective of using hot runner system is to do away with wastage of plastics material. This happens because there are no runner and sprue coming out with each shot. Since the flow of plastics is in well temperature-controlled channels inside manifold, there is much less pressure drop in the flow of plastics as compared to conventional runner system. Since there is much less pressure drop during flow of plastics, hence no excessive pressure on material is applied. Consequently, properties of plastics material are not adversely affected.

Hot runner system is generally used where multipoint injection is required *without* runner. Let us reconsider cavities and runner layout shown in Fig. 2.28. In this case, the number of hot runner system nozzle would be equal to the number of cavities.

Figure 2.30 shows a runner manifold which carry suitably designed nozzles to provide a pin point gating for each cavity. Runner manifold is kept heated by means of electric heater. There are a number of temperature sensors fitted on manifold at appropriate position. Signals from sensors are fed to multipoint temperature controller which precisely control desired temperature on all specific points. Hot runner manifold is so designed and insulated that minimum possible heat is transferred to various plates of mold. Wherever necessary adjoining mold plates are kept comparatively cool by water circulating channels.

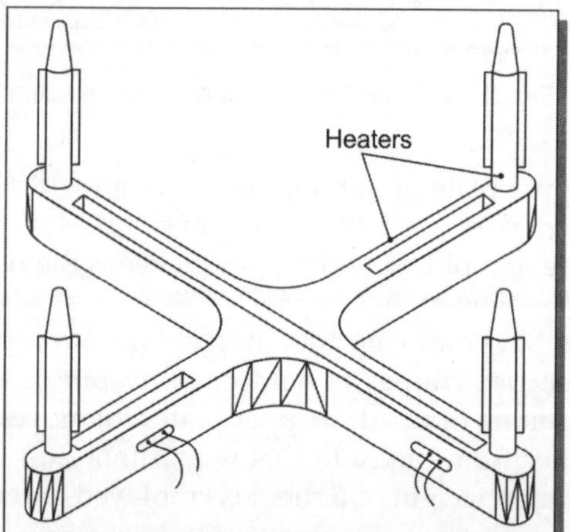

Fig. 2.30: Hot runner system nozzle

Type of Gates

Gates in injection mold cavities are small and even minute passages through which plastics enter the cavity at the time when injection of plastics takes place during complete molding cycle.

Following are the factors which determine the type and dimension of gate which may be provided in a cavity.

- Melt flow index of plastics material
- Volume of cavity
- Weight of molded components for a specific plastics/polymer
- Cavities filling speed
- Size of components. For what distance plastics has to travel inside cavity
- Overall mold temperature of runner and gates
- Heat transfer characteristics of steels in which runner, cores and cavities are made
- Component surface finish requirement
- Location of gates with respect to shape and size of component
- Plastics flow behaviour required in cavity

Figures 2.31 and 2.32 show gate sizes for same cavity but for different materials. Figure 2.31 shows typical example of gate for molding components in general purpose polystyrene. In this example width of gate is 1 to 1.5 mm and depth as 0.5 to 0.8 mm.

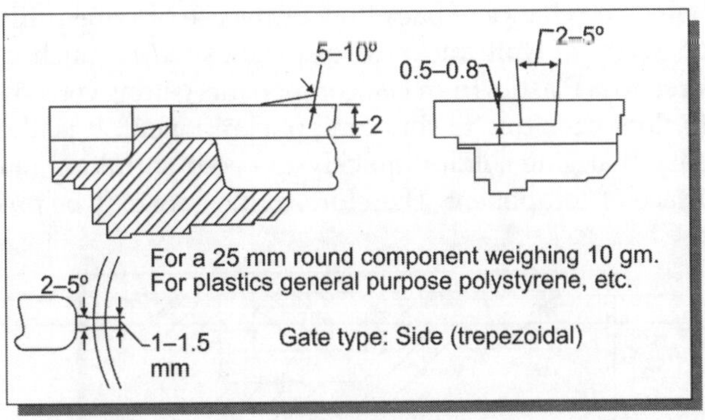

Fig. 2.31: Gate sizes

Figure 2.32 can also be a typical example of gate for polyamide 6.6. It is a syrupy plastics at processing temperature, therefore, a small size gate is needed. It may be 0.3 to 0.6 mm wide and 0.3 to 0.5 mm deep.

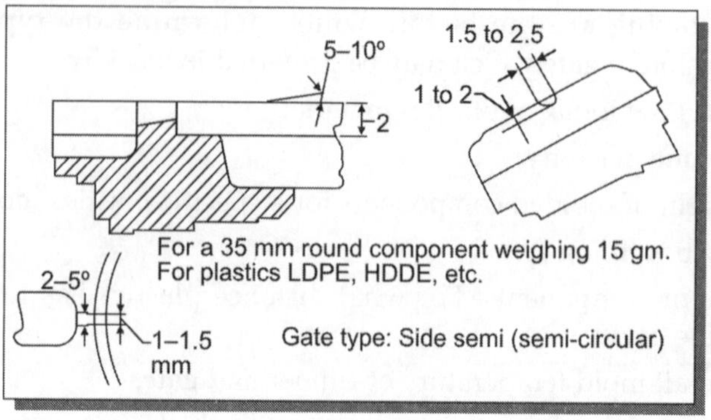

Fig. 2.32: Gate size

If the volume of cavity to be filled with plastics is confined to short dimensions as shown in Fig. 2.33a, then cross-sectional area of gate may be less as compared to Fig. 2.33b.

Referring to Fig. 2.33, volume of both the cavities is same, but the distance to be traveled by plastics in cavity '*b*' is much more and path narrow as compared to distance of plastics to travel in cavity *a*. It is, therefore, necessary that pressure drop in flow of plastics should be less to facilitate quick filling of cavity and packing. Packing means that plastics in cavity is kept under pressure till gate freezes and there is no chance of back flow or release of some pressure. In case of cavity *a*, wall thickness of plastics *Fill* is much more as compared to *b*. Plastics from *Fill* (components getting cooled) would tend to flow back due to shrinking of component. It is, therefore, necessary that gate solidify quickly to avoid shrinking marks on the surface of component. Therefore, a fine gate may be provided.

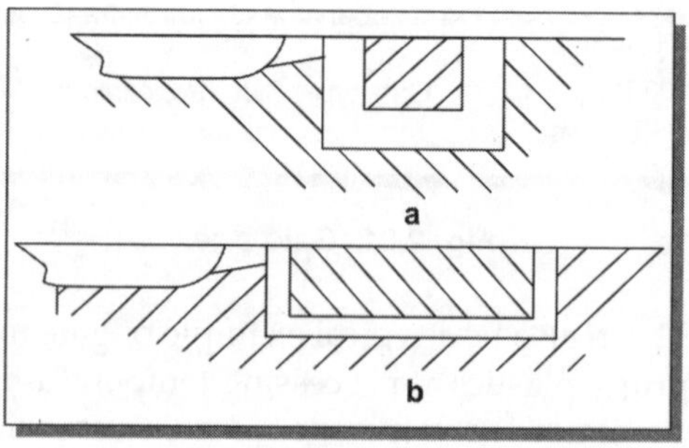

Fig. 2.33: Gate size cross-section

Shape of gate may be semi-circular or trapezoidal as shown in Fig. 2.32 and 2.31, respectively.

In Fig. 2.34, dimensions of cross-section are shown. On calculating cross-section area of gate, it is found that for same width and depth, area of trapezoidal cross-section is more. So it is preferred, if less pressure drop in plastics flow across gate is required.

Entry of plastics into cavity with very high speed, resembling to a jet of plastics material. This, many a times, causes flow marks on the surface of components. This may be a visual quality deficiency. In some cases, it may be functional as well.

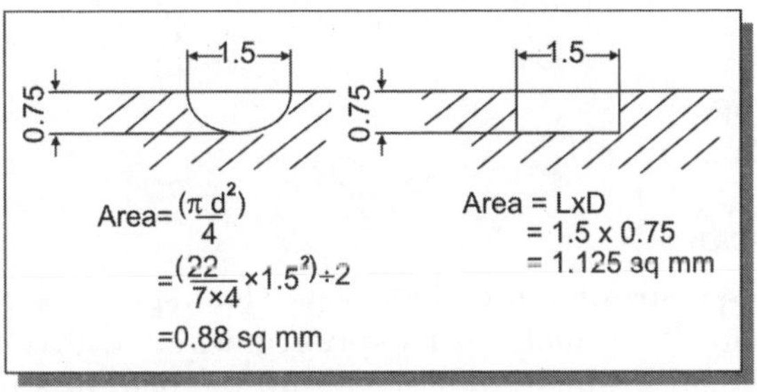

Fig. 2.34: Dimension and gate cross-section

Figure 2.35 shows a clear PMMA (polyethylene methacrylate) lens together with its runner and gate.

Figure 2.35a has a side gate. Lens molded with this gate has a flow mark which is highly undesirable. It could not be completely eliminated by varying processing temperature, pressure and speed.

Flow marks could be eliminated by using a *fan gate*, which has exit width more than entry width. Consequently, jetting action gets reduced and components (lens) free from flow marks and weld lines are obtained. Figure shows a lens with fan gate.

Following are the types of gates which are generally provided in cavities.

- Direct
- Side
- Disc

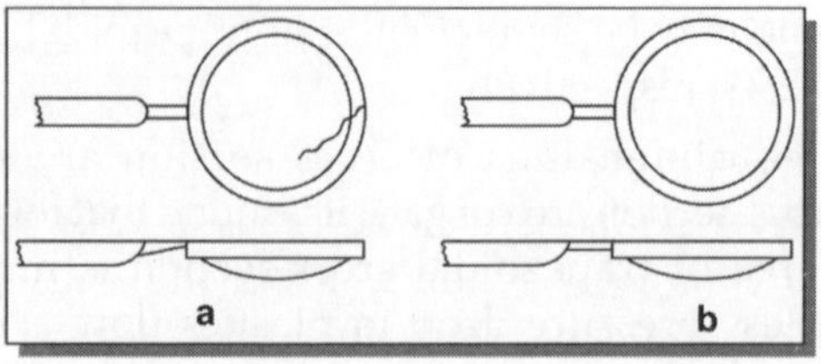

Fig. 2.35: Weld line (*see* Photo 2.8)

- Pinpoint
- Submarine
- Tab
- Fan
- Film
- Hot tip
- Banana

Direct Gate

Figure 2.24 is an example of direct gate. Here, sprue directly opens in the cavity. When mold opens, sprue comes out with component. Sprue is then manually removed with the help of a jig cutter. Diameter of sprue hole in the cavity is generally kept 50 to 60% of part's wall thickness at the sprue point. If wall thickness is found to be too less for a sprue diameter, the wall thickness is increased at the point of sprue by provided dome-shaped small space in the core and hence component takes that shape (Fig. 2.36).

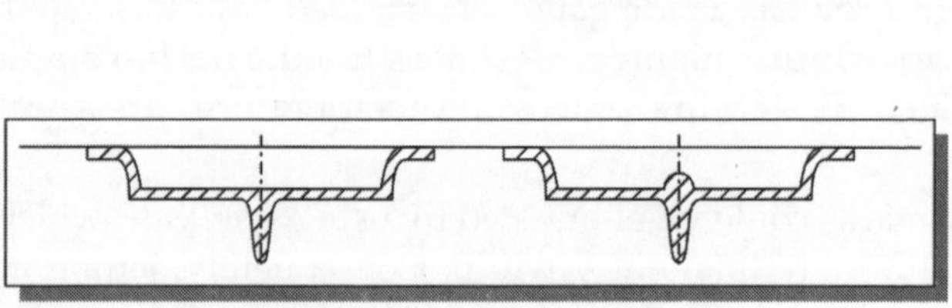

Fig. 2.36: Cole well

Side Gate

Side gate is mostly used in a variety of cavity design and polymer being molded. Figures 2.31 and 2.32 give some idea of side gate design.

Disc Gate

Disc gate is used for the components where no weld joint is tolerable. Whether it is visible or not. It is generally used for cylindrical components which are open on both the ends. Figure 2.37 shows a component with disc gate attached.

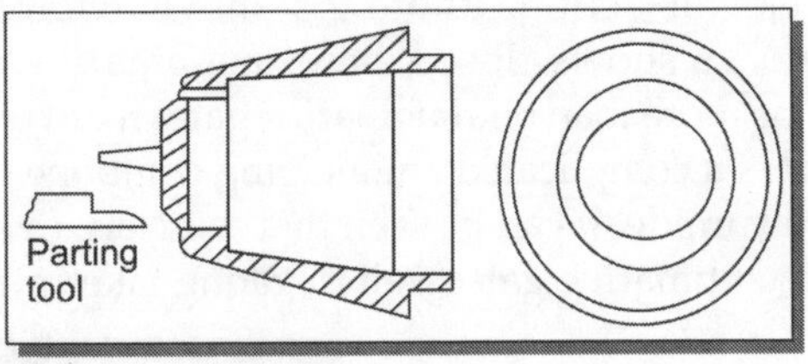

Fig. 2.37: Disc gate

A secondary operation is needed to remove the gate by a parting tool on a lathe. A jig may be loaded on lathe journal. Component with gate is placed on jig to rotate. While rotating, gate is removed by penetrating a parting tool at an appropriate location.

Pinpoint Gate

Hot nozzle of injection cylinder remain in touch with a small pin hole in the base of cavity which is very short in length, say from 0.8 to 1 millimeter. During mold opening, stroke component moves out of cavity together with core. Gate gets detached from nozzle orifice.

Photo 2.4 shows possibility of having more than one cavity. In this case, a 3-plate mold is used having a non-hot runner.

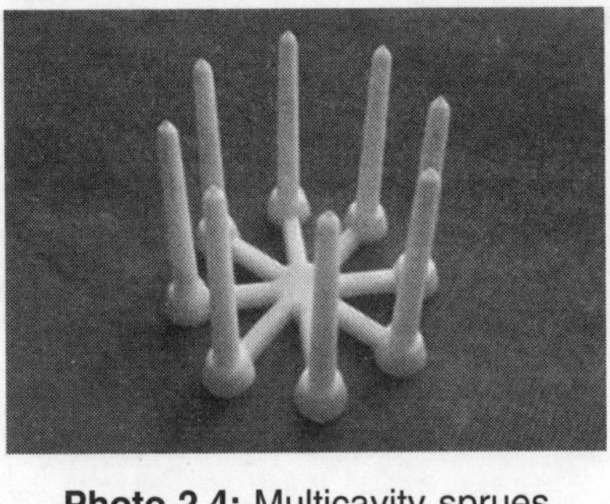

Photo 2.4: Multicavity sprues

Submarine Gate

Figure 2.39 shows a general design of submarine gate. There in a small pin hole connection between cavity and runner, end of which is machined to reach near the cavity wall. Mould temperature and cycle timing is kept in such a way that angled end of runner which is still in pliable condition comes out at the time of mold opening or ejector action. Photo 2.5 shows complete runner portion. Photo 2.6 shows closed view of submarine runner end.

There may be more than one submarine gates for even one cavity. Photo 2.7 shows a complicated engineering component molded in glass filled polyamide. It can be seen that there are five submarine gate points. In submarine gates, self-degating takes place.

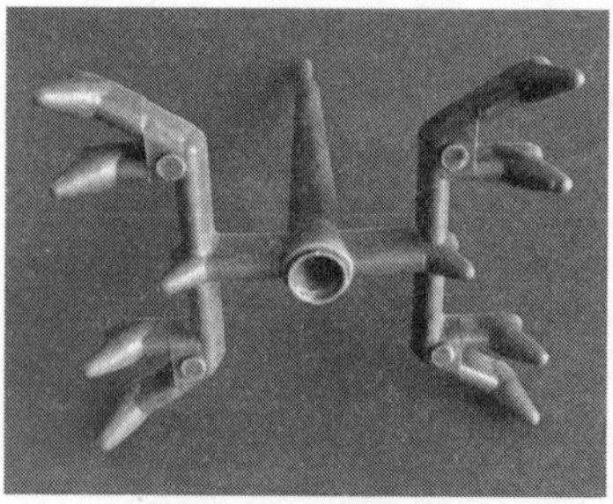

Photo 2.5: Submarine runner

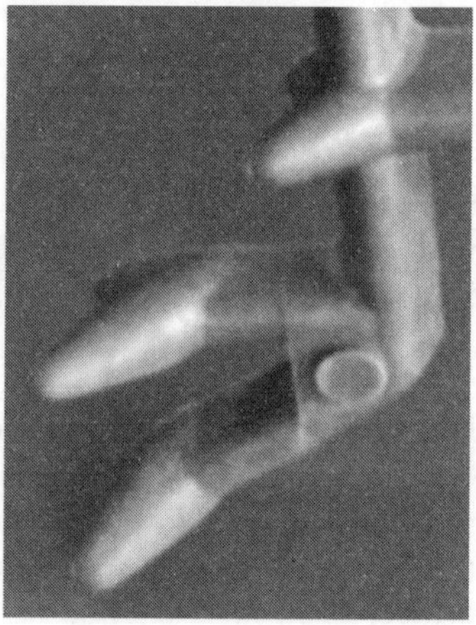

Photo 2.6: Close view of submarine gate

Photo 2.7: Complicated engineering component

There are instances where gate of any type on out side surface of part is not permissible for cosmetic reasons. In such cases, a tab is provided in the back of component or part. Plastics is injected by means of submarine gate. This arrangement is show in Fig. 2.38.

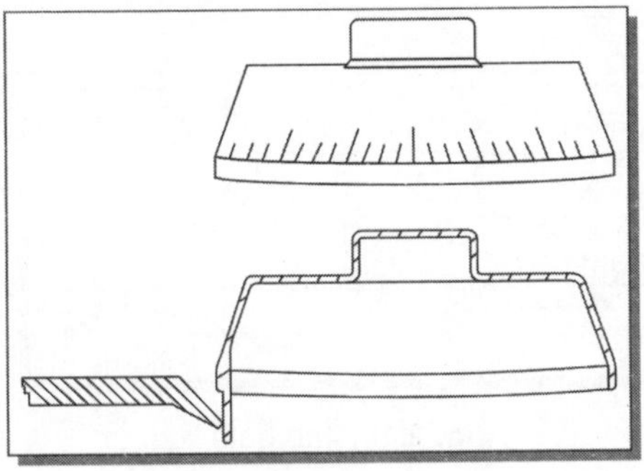

Fig. 2.38. Submarine gate

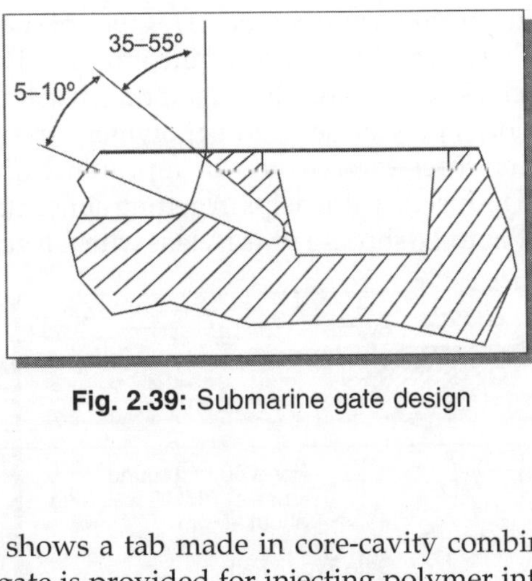

Fig. 2.39: Submarine gate design

Tab Gate

Figure 2.40 shows a tab made in core-cavity combination and a submarine gate is provided for injecting polymer into cavity.

Tab may be removed by a suitable secondary operation cutting or hot shearing.

Figure 2.40b shows a disc which should be completely free from flow lines which is on many instances difficult to avoid with a side gate. This is due to fast jetting of plastics into the cavity. This difficulty may be overcome by adopting a tab gate as shown in Fig. 2.40b. Disadvantage with tab gate is that it needs a secondary operation to remove it. There is likelyhood of getting surface finish scar at the point of gate removal. It would not pose a problem, if brim of disc is hidden in assembly together with other components.

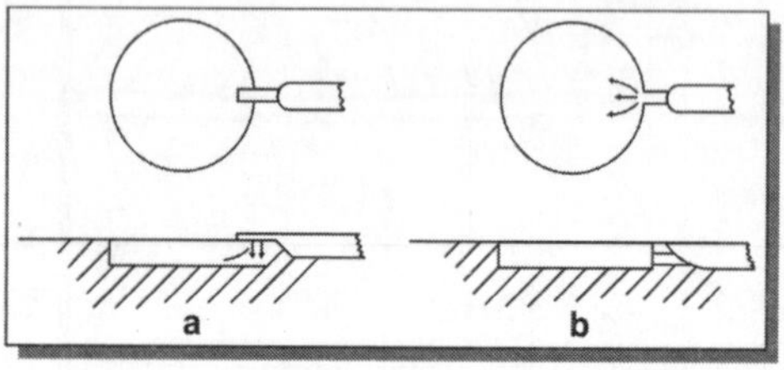

Fig. 2.40: Tab gate

Fan Gate

Figure 2.41 shows a fan gate. Gate end meeting runner is narrow.

The other end of gate is much wider. This type of shape, which is called *fan gate,* reduces the speed of incoming polymer due to expansion of cross-section area. Further more, instead of having a direct high velocity polymer jet, flow of polymer is in almost all 150 degree direction. Therefore, chances of appearance of flow marks are very much reduced. It also helps in getting abnormal stress free components. Photo 2.8 shows a PMMA lens with fan gate attached.

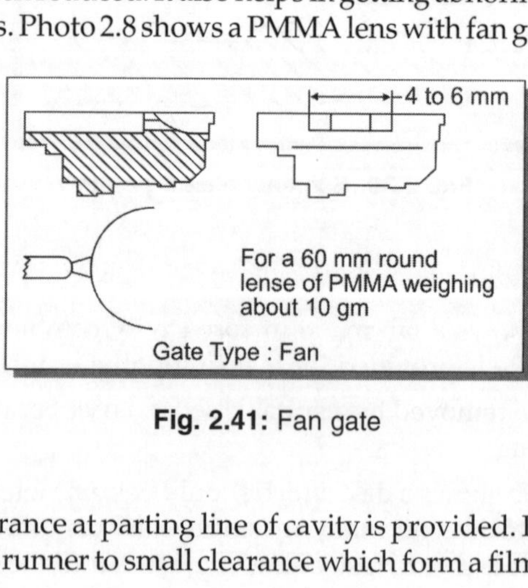

Fig. 2.41: Fan gate

Film Gate

A small clearance at parting line of cavity is provided. Polymer first travels from runner to small clearance which form a film. Full length

Photo 2.8: Component with fan gate

of film open to cavity injects plastics into cavity. Such a shot is shown in Fig 2.42.

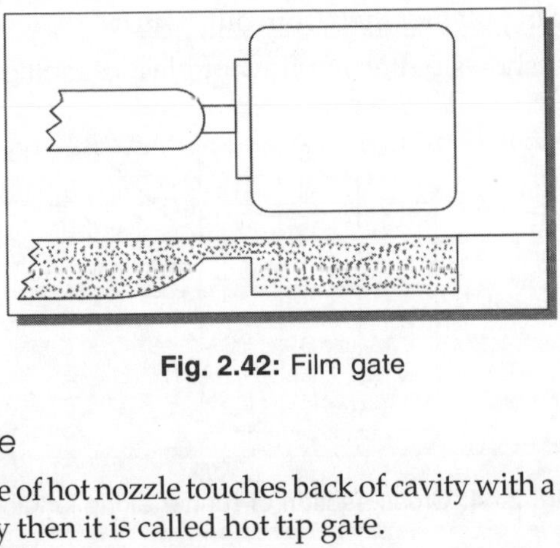

Fig. 2.42: Film gate

Hot Tip Gate

When orifice of hot nozzle touches back of cavity with a fine opening in the cavity then it is called hot tip gate.

Banana Gate

Figure 2.43 shows shape of a banana gate.

This gate is machined on mold parting surface. Runner is shaped in such a way that gate is located at the base of component. Name *Banana* is driven from its shape.

Flow of Plastics into Sprue, Runner, Gates and Cavities

Polymer melt enters runner to travel to fill cavities. As soon as melt enters the runner, heat from outer surface of flowing plastics

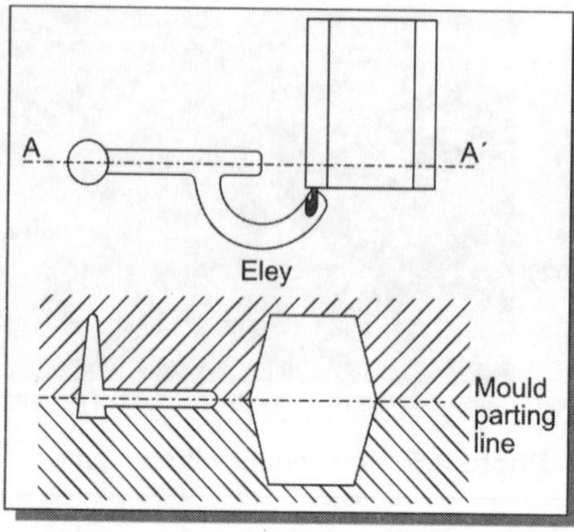

Fig. 2.43: Banana gate and cross-section of runner

looses to cool steel walls of runner. This results in progressive solidification of polymer melt from outer surface to inner core.

Figure 2.44 shows pattern of flow of plastics melt.

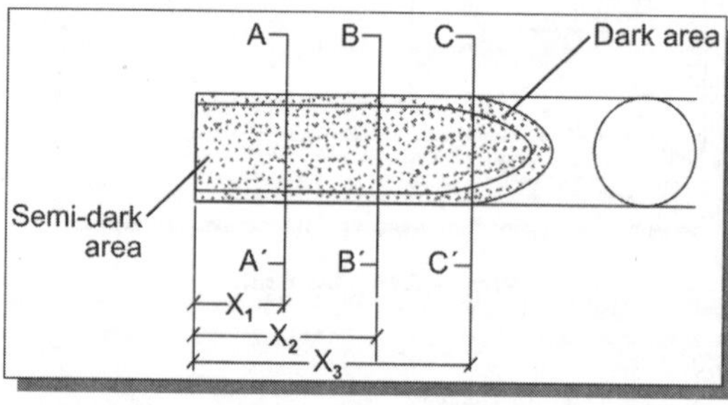

Fig. 2.44: Cross-section of runner along length

Dark area in figure shows plastics which has become semi-solid or solid due to transfer of heat to steel in which runner is made. Section views at AA′, BB′, and CC′ are shown in Fig. 2.45.

These sections are taken at distances X_1, X_2, and X_3 from entry point of runner.

It is evident from Fig. 2.45 that cross-section of melt keep on getting reduced with the advancement of flowing melt.

Assuming that volume displacement rate from injection cylinder is 50 cc per second then plastics flow speed in a runner of 5 mm

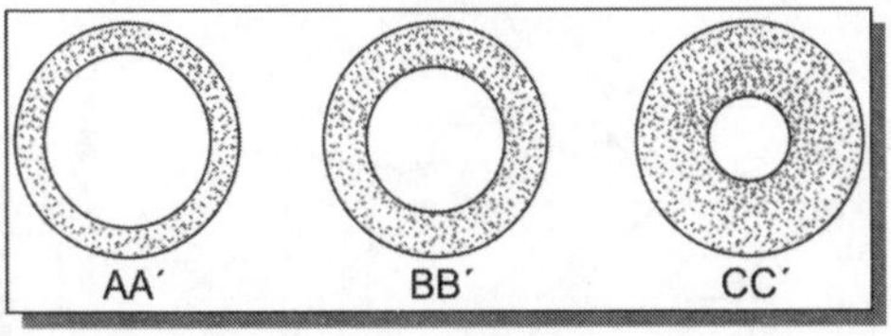

Fig. 2.45: Sectional view of flow at different places

diameter would be 285 cm per second. In practice, flow of plastics cannot cover a distance of 285 cm as melt is not flowing freely. It is experiencing resistance in flow due to cooling down of melt and other factors like flow properties of polymer, surface polish of runner and air exhaust passage in cavities.

Referring to Fig. 2.45, cross-sectional view CC′ is almost the limit for leading end of polymer flow to reach cavity gate because still there is molten core.

If injection pressure and speed is increased, section BB′ (Fig. 2.45) can be achieved near the cavity gate. A better cavity filling can then be achieved.

Increased mold temperature, specially along the runner system may cause reduction in injection pressure and better cavity filling.

On the other hand, increase in mold temperature may require cooling time for complete shot longer to attain a stable condition and there is no chance of sacrificing component quality.

Various designs of gates are already discussed. Purpose of particular design is also mentioned. Now some details about gates are mentioned here. Basic function of cavity gate is to provide passage for polymer melt to enter into cavity in such a way that component free from molding defects are obtained. Further, a suitable gate section design with minimum possible area should be the aim without sacrificing component quality. Excessive injection pressure may also be avoided. All these considerations point to the necessity of having a compromise between variables. Figure 2.46 shows following dimensions of gate with respect to runner

X = Depth of gate opening towards cavity
X_1 = Opening of gate towards cavity
X_2 = Entry opening of gate towards runner

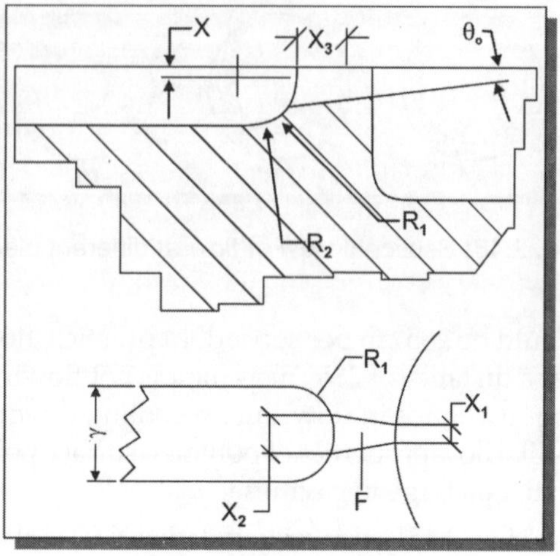

Fig. 2.46: Gate design

X_3 = Length of gate that is steel thickness between cavity and runner walls
R_1 = Radius at meeting point between X_1 and X
R_2 = Radius of runner at the end where gate meets
R_3 = Radius at gate and runner meeting point
F = Finish of all surfaces of gate
Q = Slope of gate land

Values of above dimensions may be as follows for a cavity 5 cm in diameter and 0.6 cm deep for a disc to be molded in polypropylene

X = 1

X_1 = 1.3 mm	X_2 = 2.2 mm
X_3 = 2.0 mm	Y = 4 mm
R_1 = 0.5 mm	R_2 = 1.3 mm
R_3 = 0.5 mm	F = very smooth and polished

In practice, gate dimensions are kept a little smaller. After first trial of mold, if necessary, dimensions are increased for desirable molding quality.

Flow of Polymer Melt into Runner, Gate and Cavity

Ultimate objective is that all the cavities of a multicavity mold should get filled up with similarity. This ensures uniformity in quality of components. Figure 2.47 shows balanced layout of runner. Balanced layout means that polymer melt reaches all cavity simultaneously with similar melt condition, speed and pressure.

To examine flow behavior of melt, incomplete filled shots may be taken out. This is shown in Fig. 2.47. Cavities towards centre of mold are not completely filled. Whereas cavities towards outer side of mold are completely filled. Why is it so? It is explained with the help of Fig. 2.48.

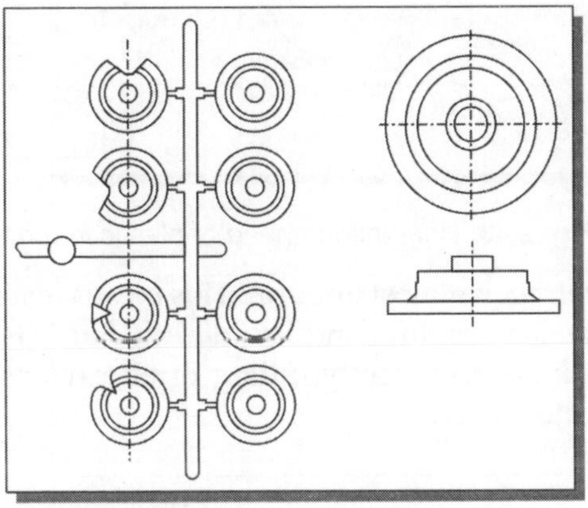

Fig. 2.47: Balanced layout of runner

Melt flowing in primary runner have less fluid skin near the wall of runner. Melt in the centre of runner is more fluid (Fig. 2.48). As soon as melt reaches secondary runner, flow of melt is divided into two directions, as shown. On account of flow pattern, melt in secondary runner towards mold centre is less fluid as compare to melt towards outside of cavities. This results in non-uniform filling of cavities. This effect is shown in Fig. 2.48.

Imbalance insides of flowing melt may be improved to a great extent, if shape of runner is modified as shown in Fig. 2.49.

A *cold well* is made on secondary runner in front of primary runner. Instead of taking turns, leading end of melt enters a space, generally known as *cold well*. As soon as melt flow meets a stop in cold well, flowing melt *burst* outside of less fluid layer to secondary runner as shown in Fig. 2.49. It is shown that flow pattern in

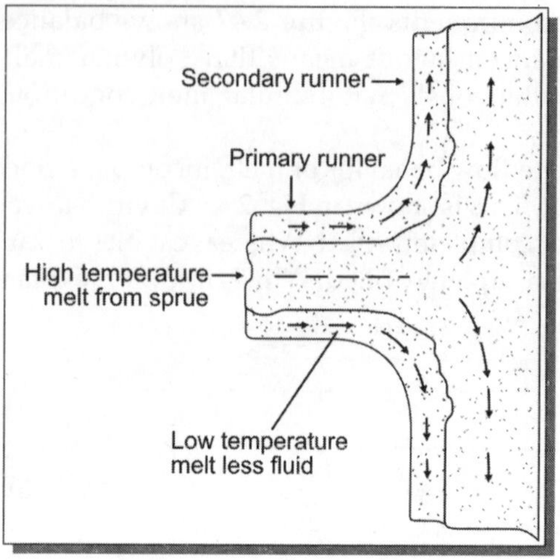

Fig. 2.48: Non-uniform head of plastic in runner

secondary runner is almost uniform. This means that flowing melt in the centre is more fluid and having less fluid layer around it symmetrically. Consequently, filling of all cavities takes place almost simultaneously.

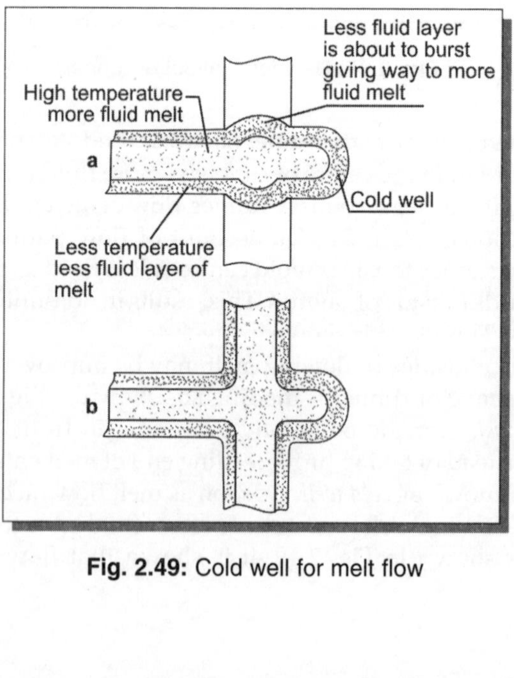

Fig. 2.49: Cold well for melt flow

Multicolor Injection Molding

Object of molding a component in two colors is to have a better visual effect. Further, two colors remain permanent. Key of a keyboard of electronic typewrite or computer may have alphabets printed on key surface. Printed alphabets or numbers may become dull or disappear all together with regular use. If alphabets are molded in the body of key in different color, say white then white alphabet with black key body may be a permanent combination. Two-color molding is quite expensive as compared to single color molding. Still it is sometimes done in case of sophisticated components.

Two-color molding is normally done with similar polymer. Color may be different. Two-color molding may be done in two ways:

1. One part molded in a color. It is then used as insert and inserted manually for *insert* molding.
2. Another option is that a rotary type mold may be used on a machine which has two plasticizing cylinders with separate nozzles, to meet two different feed heads of the mold. Referring to Fig. 2.50, plasticizer P_1 is used for white ABS and plasticizer P_2 is meant for black ABS.

Another method of double color molding is simultaneous injection of two different color polymers of same type and grade. A peculiar two color designed surface is obtained. Flow of two different color

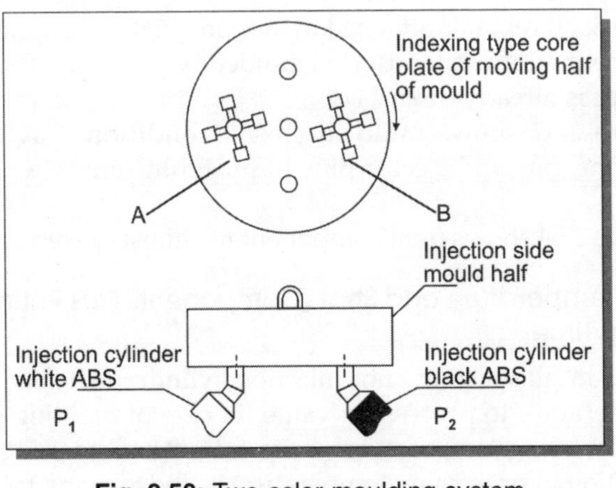

Fig. 2.50: Two-color moulding system

plastics produces surface of varied pattern design. For every shot, design is different due to flow behaviour of melt.

Sequence of operation may be as follows:

- Cavity cluster 'A' is towards left hand
- Mould closes
- Only left hand side injection cylinder injects white ABS.
- Right side injection cylinder having black ABS remains inactive.
- Mould opens. Submarine gate snaps from component and runner comes with moving side of mold. Runner is ejected and components remain on cores.
- Core plate of moving side of mold gets indexed in clockwise direction by 180°.
- Mould closes. Now cores with white molded parts enter cavities. Parts loaded on cores are inside cavities as inserts.
- Both left and right side injection cylinders inject ABS in both cluster cavities. Cavity cluster 'B' is now on left side and 'A' on right side.
- 'B' is filled with white ABS while in 'A' *insert* molding is done with black ABS
- Mould opens
- Ejection of two-color molded components and runner take place. In this way, molding continues to take place.

Figure 2.51 shows typical design of insert which gets molded in white ABS. It remains attached to the core. Figure 2.52 shows final component after white insert is molded with black ABS. Process sequence is already described.

Figure 2.52 shows mold in closed condition. Cavity has a submarine gate and ejector pins to push out runner when mold opens.

Figure 2.54 shows final component in almost ejected condition.

Mould Temperature and Shot (Components Plus Runner) and Cooling Time

Polymer in plasticizing cum injection cylinder is heated to raise its temperature to processing value. In case of molding grade of ABS, processing temperature varies from 246°C to 287°C. Range of variation in processing temperature is due to many factors like

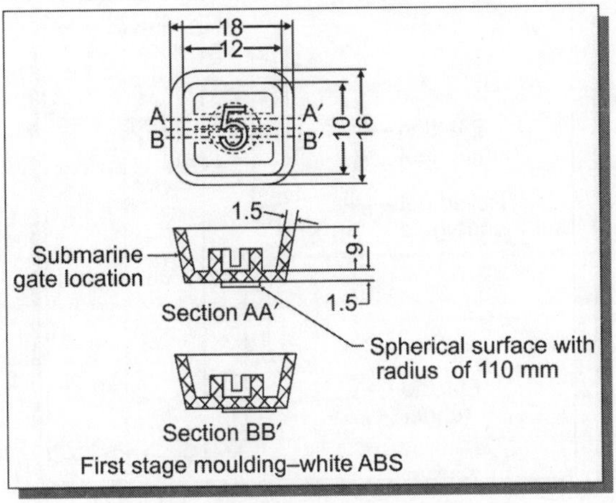

Fig. 2.51: Multicolor molding

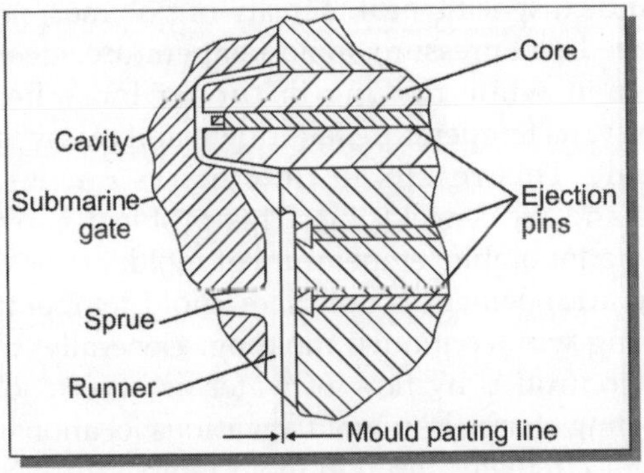

Fig. 2.52: Ejection and submarine gate

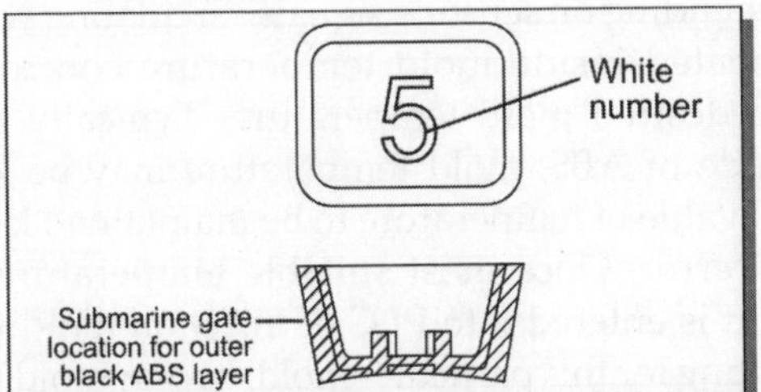

Fig. 2.53: Two-colour molded components

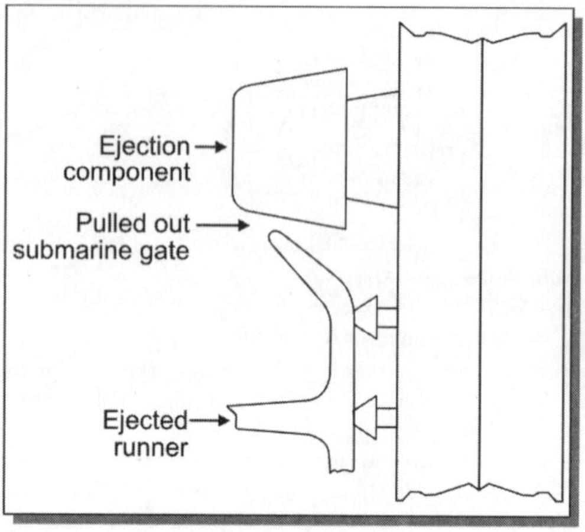

Fig. 2.54: Ejector system

melt flow index, specific heat, density of polymer, mold design, injection speed and pressure, mold temperature, etc.

Plastics melt while flowing in runner loose heat to steel. Consequently, its temperature fells down and cavity is filled with less hot melt. This results in flow marks on the surface of component and weak weld line. This problem is overcome by maintaining a favorable temperature of mold.

If cooling arrangement is not there, mold temperature would keep on rising with continued molding. Generally, temperature of mold is controlled by flow of water or other cooling media through cooling channels in mold at various locations. Generally, cooling water circulates through mold temperature control unit. Circulating water is cooled or heated, depending on temperature of mold. A temperature sensor is fitted to mold at an appropriate location. Depending on sensor's signals, circulating water is either cooled or heated inside mold temperature control unit, thus maintaining desired mold temperature. Typically for injection molding grade of ABS, mold temperature may be kept around 60°C. Actual value of temperature to be maintained is determined by trail and error. Once most suitable temperature of mold is established, it is entered into PLC of injection molding machine, for future setting for this particular mold. In case mold temperature is more than required, components may have sink marks and reduced dimensions. Mold stay time might have to be increased to reduce sink marks and maintain specified dimensions of components.

In case mold temperature is less than desired value then the chances are that components might have flow marks and poor weld lines. There can be de-gating problems for submarine gates. There are some mold designs where HDPE component having shallow threads are *jump* ejected. This is possible, when HDPE is in semi-solid state. In case temperature of molded component which is still on core gets reduced too much than *jump* ejection would not take place successfully. There would be chances of distortion in thread profile or failure at weld line.

Mould Closing and Opening

When plastics melt reaches processing temperature, mold is closed for injection. Closing of mold by manual control is now almost obsolete. Complete molding cycle takes place automatically in modern injection molding machines. Data like sequence of operations, temperature and timing for various stages of molding cycle for a particular mold are saved in PLC of machine. Programmable logic control takes care of mold closing, opening and ejector action, etc.

Many a times mold closing and opening is in stages. This depends on design of components and mold. It is explained with the help of Fig. 2.55 and photo 2.9 of component.

Photo 2.9 is of a flashlight body molded in polypropylene co-polymer.

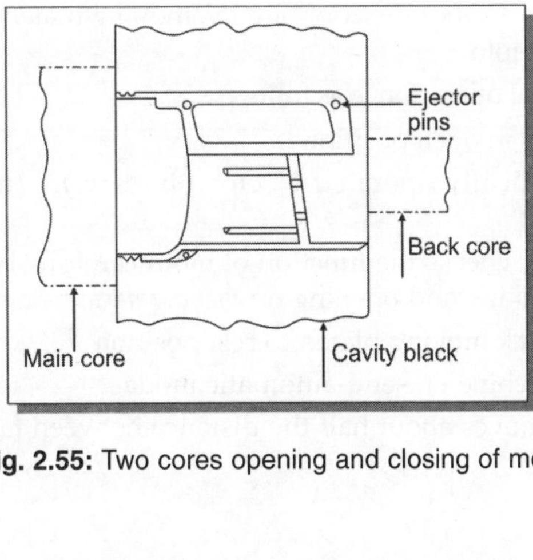

Fig. 2.55: Two cores opening and closing of mold

Photo 2.9: Flashlight body

Figure 2.55 is a schematic sectional view of flashlight body inside moving half of the mold. There are two moving cores. One is main core and other is back core. They move in and out at right angle to ejector pins. Sequence of operation of mold closing and opening should be such that cores and ejector pins do not foul. Mould closes completely, when cores are in locking position in the mold halves when closed.

Author of this book has designed component and mold. A fully automatic 450 tons locking force machine is chosen to run the mold. Machine is microprocessor controlled. It also has facility to set mold closing in two steps. This means that moving platen with mold half moves for a pre-set distance then it stops for cores to take *IN* position. Cores move by hydraulic jacks mounted on moving platen. Once the cores are *IN*, moving platen starts again to close the mold.

Sequence of operation is as follows:

- Mould is in open position
- Hydraulically operated ejector pins are *IN* (non-ejection position)
- Operator checks the function of mold carefully by operating mold locking and opening on *inching mode*
- Bring back moving platen to rest position
- Start machine on semi-automatic mode
- Platen moves about half the distance between faces of mold halves

- And stops there
- Cores move automatically to their *IN* position
- Platen starts again to close the mold with low force locking and then
- Immediately full locking force is applied, if no obstruction is sensed by machine
- Injection take place
- Cooling time is given
- Mould opens up to some distance, say three-fourths of stroke
- Cores are automatically withdrawn to *OUT* position
- Hydraulic ejector system pushes the ejector pins of mold out
- This action ejects the component out of mold
- Component fells on conveyor and taken away from machine
- Once satisfied, operator switches the machine on fully automatic mode
- Robotic system may also be used for pick and place action

Molding Cycle Sequence

In case of non-insert molding, sequence of molding operation may be as follows.

With pre-set timer heating of plasticizing cylinder, start of mold temperature controller and water circulation starts. Timing may be set in such a way that machine becomes ready to start by the beginning of production shift.

- Operator check the setting of parameters on microprocessor (PLC)
- Drools a little amount of molten plastics by manually *inching* injection screw
- Removes drooled plastics (say about 10 to 30 grams)
- Checks closing and opening of mold on *inching mode* for proper functioning
- Press button on PLC for semi-automatic molding cycle
- Mould starts to close
- Mould closes and full locking force is applied
- Injection carriage has already travelled toward mold feed head
- Nozzle touches mold feed head with pre-set force
- Injection of plasticized plastic take place till mold cavities and runner are completely filled
- Mould cooling time start

- While cooling still taking place and about to be completed, injection screw plunger is withdrawn a little bit so that no drooling of melt take place
- Injection carriage moves back.
- Mould opens
- Robotic picker comes in between the two-halves of mold, ready to hold molding as soon as it is ejected by ejector pins or sleeves
- Robot pick up goes out from mold opening together with molding
- Mould closes again to repeat the cycle
- While cycle is going on, robot places the molding at targeted location and gets ready for next *pick up*.

In case of insert molding, robot picks up inserts and load them in the mold. As soon as robot pick up head goes away from in between the mold halves, mold start closing and complete cycle starts taking place.

Ejection System

After completion of molding, cycle, molded part has to be ejected from cavity or core. Ejection may be affected. Following two aspects of ejection system are taken into consideration:

1. Ejection system *built up* in mold
2. Ejectors actuation system in the machine

Ejector system *built up* in molds vary from design to design of mold. If mold has a single cavity with sprue hole directly opening in cavity then after molding, component with sprue moves out of cavity or core of moving half of mold. At the end of opening stroke hydraulically operated pusher rod pushes ejector pins out to a certain length which ejects the component out of core. Figure 2.56 shows this arrangement.

Core (1) is a stationary part of the mold. Component (2) is shown on the core before ejection. Part (3) is a sleeve ejector around core (1). In back position (non-ejection position), face of ejector sleeve is in line with face of plate (4) which meets face of cavity plate towards injection side. (5) are three push rods attached to sleeve plate (7) which is outside the mold on which three rods are attached. Gap between plate (7) and back plate of mold is enough to accommodate springs. Block (8) ensures that spring coils are not crushed. (9) is

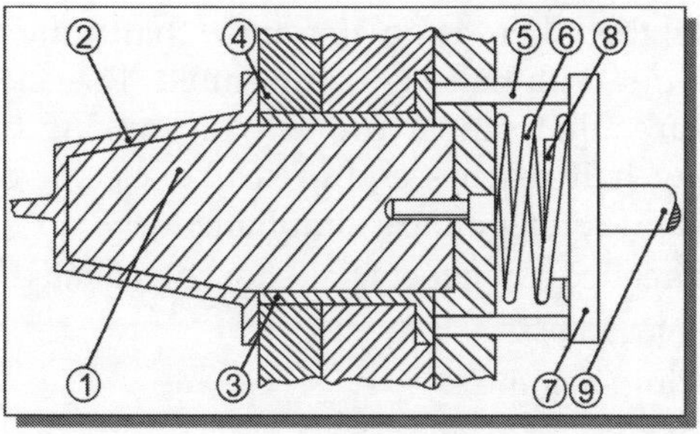

Fig. 2.56: Ejection system

the rod from hydraulic jack. As soon as ejector actuation command is given, rod pushes the ejector plate. In turn, three push rods push the sleeve out, while sliding over the core. Forward motion of sleeve ejects the component. Suitable undercut, beads or roughness is provided on core so that component definitely comes with core while mold is opening.

Another possibility of ejection is illustrated in Fig. 2.57 where components and runner has to be ejected.

A section of mold in closed condition is shown. Cavities and runner have injected polymer which is getting cool to attain a temperature suitable for ejection. (1) is moving side of mold and (2) is injection side half of mold. In case holding taper recesses (7)

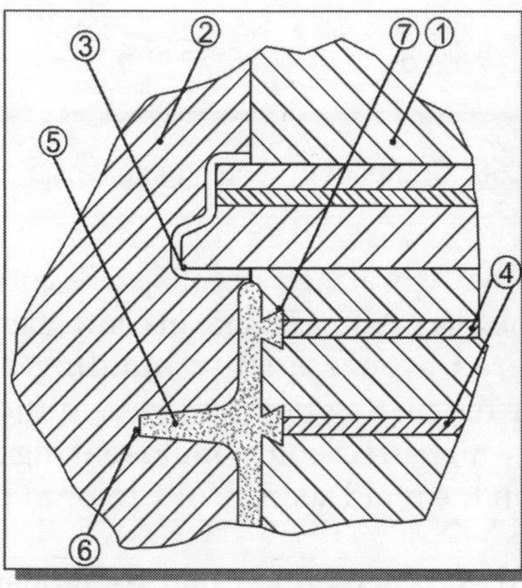

Fig. 2.57: Runner pulling and ejection

are not made, then there is no guarantee that runner with sprue would come out with moving half of mold. Presence of putting recesses ensures that runner comes with moving half of mold. When moving half reaches end of stroke, ejector pins (4) eject components and runner almost simultaneously.

In case of threaded components, ejection is facilitated by means of:

- Collapsible cores
- Unscrewing system

Choice of above alternative ejection system depends on design of component. Figure 2.58 shows a component of plating grade ABS.

Due to esthetic reasons, no cut or recess can be provided to unscrewing the components, so a collapsible core system can be adopted. But with collapsible core system, there are chances of flashes near the threads where wedge meet core surface.

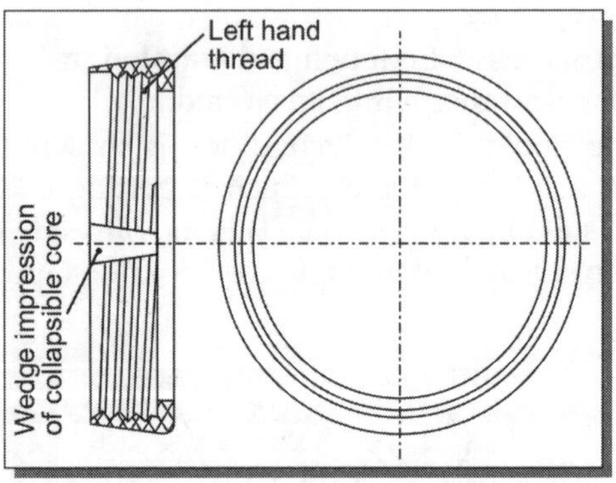

Fig. 2.58: Component requiring collapsible core

Figure 2.59 shows typical view of collapsible core. Area (f) along joining faces is the place where flashes in component may develop. This is considered a poor quality molding as screwing of component over the mating part would be difficult or may not screw at all. Very high precision in mold making and cleanliness is required to run the mold successfully to produce components free from flashes.

Most of the times, it is required that component should have thread without any break so that there is no question of thread break, therefore, unscrewing ejection. Figure 2.60 shows a high impact polystrene component for which mold would have to be made with unscrewing system.

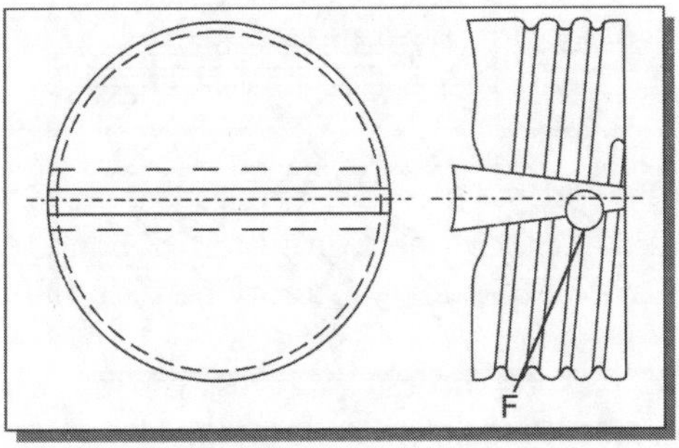

Fig. 2.59: Core with taper wedge

Component has external thread (3) and internal thread (2). Four sunken notches (1) and raised notches (4) are provided to facilitate rotation of component in anticlockwise direction. Component moves out of thread core while it is rotated.

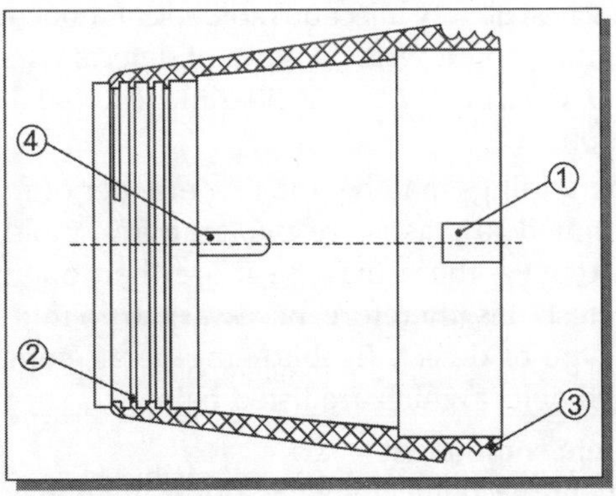

Fig. 2.60: Two-side threaded component

Basic construction of core cavity design of a 2 cavity mold is shown in Fig. 2.61. (1) is fixed core with cooling water channel (5). (2) is cavity space. (3) is rotating sleeve with sunk and raised notches. Mould opens completely, sleeve starts rotating in anticlockwise direction. Since both external and internal threads in component are 2.5 mm pitch right hand threads, component

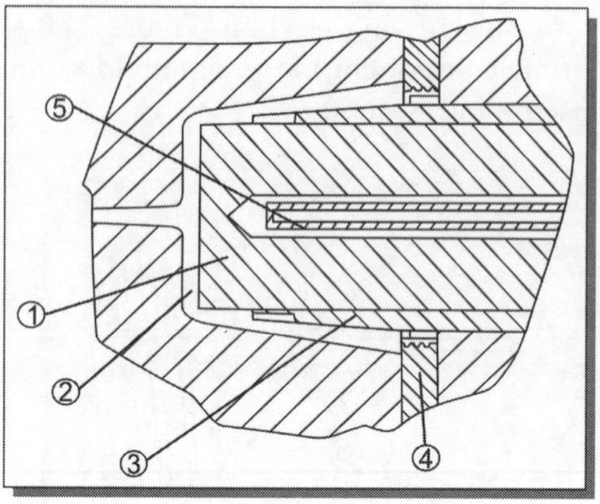

Fig. 2.61: Progressive filling and weld mark

moves out of core and sleeve axially. At the end of ejection cycle, rotation of sleeve stops. Components are then allowed to either fell down on conveyor or picked up by a robotic picker.

Defect in Molding, Cause and Remedies

For producing best quality injection molded components, it is almost essential to know about various types of defects which are found on the components, only then it is possible to avoid them and take remedial action.

Defects in moldings may be due to a number of reasons which may be attributed to plastics resins/granules, mold or molding process. Moreover, there may be more than one reason for a particular defect. It is, therefore, necessary to go into some details about each type of defect. In following paras, some details are given for each defect which are listed below.

- Incomplete component
- Component mechanically weak opposite to injection point
- Weld line visible
- Flow lines on surface of component
- Silver streaks
- Black spot
- Sink marks
- Flashes
- De-gating not smooth

- Ejector mark
- Any other marks
- Clarity below standard
- Microdots on surface
- Distortion in profile

Incomplete Component

Incomplete component means that cavity is not completely filled and packed with plastics melt. Most likely cause is that desired quantity of plastics is not injected. Another reason may be that mold is too cold, as a result melt gets solidified before cavity is filled up completely. Low injection pressure may also be the cause of slow filling of cavity. Other reason may be that gate area is too less to allow plastic melt to flow with speed. First of all corrective step may be re-setting of weight/volume of injected plastics, injection speed and mold temperature. Increase in gate size may be resorted to in the last, if other adjustments do not give desired result.

Component Mechanically Weak

Figure 2.62 shows flow pattern of melt in cavity. On entry into the cavity, melt flow bifurcates into two directions as shown. When two portions of flowing melt meet, they get fused to each other provided temperature, viscosity and pressure are favorable. In case variable conditions are not favorable then melt flow from two directions will not fuse properly. This results in a weak weld line which may not be visible by naked eye.

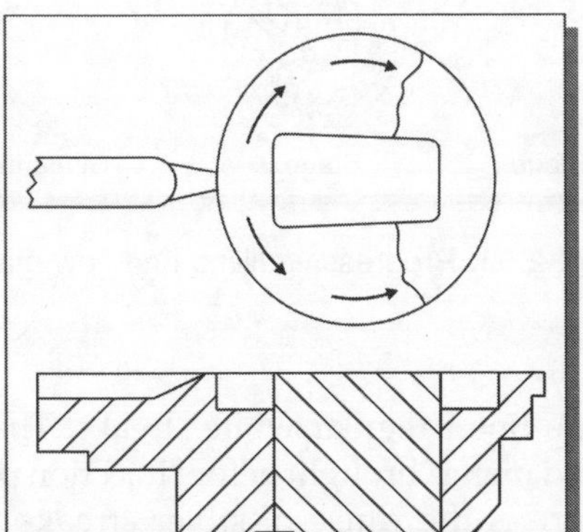

Fig. 2.62: Progressive filling and weld mark

Weld Line Visible

Sometimes in spite of suitable processing conditions, weld line remains visible. Cause of visibility of weld line may be rough surface of cavity. Smooth and polished surface of cavity very much reduces chance of weld line becoming visible.

Flow Lines

Flow lines on the surface of components are very fine lines which may or may not be visible by necked eye. Unlike weld lines, flow lines are only on the surface and not throughout the wall thickness. Generally, flow lines are caused due to cavity shapes, configuration and slow speed of injection.

Figure 2.63 shows three stages of cavity filling with low speed. Stage I shows that injected plastics is taking shape of strand. Stage II shows that cavity is fully filled with strand but full injection pressure has not yet developed.

In stage III, plastics melt is completely packed but strands which have become more solid cannot be fused together and thus few lines remain visible. These lines are called flow lines. These can be eliminated by increasing injection speed and mold temperature.

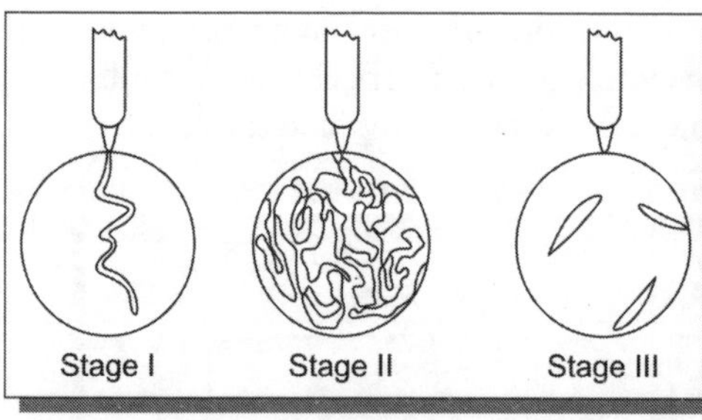

Fig. 2.63: Progressive filling and flow mark

Silver Streaks

Silver streaks are fine irregular white streaks. These are formed when over heated melt is brought before injection piston. Trapped gas and moisture is the cause of silver streaks. This defect is generally avoided by drying plastics granules before it enters, throat of plasticizing barrel, controlling melt temperature precisely and avoid long residence of polymer in front of injection head of plasticizing screw.

Black Spot

When mold is closed, space between cavity and core is closed. Generally, mating surfaces of two halves of mold are not absolutely matching. Hence, there is always a chance for air inside the space to escape, when plastics melt enters cavity.

But sometimes mating surfaces are so polished and well matched that there remains no space or passage of air to escape. In such a cases trapped air, due to compression, becomes so hot that it burns plastics melt in its vicinity. So a black spot is formed on the component.

This defect can be avoided by providing very fine passage for air to escape. But dimension of air vent should be so small that only air can escape and plastics melt cannot flow.

Location of air vent is normally chosen where weld lines are expected to occur. If air volume to escape is large, then a larger area of air vent is needed. In such a case, an air vent with minute depth but longer side may be the solution (Fig. 2.64).

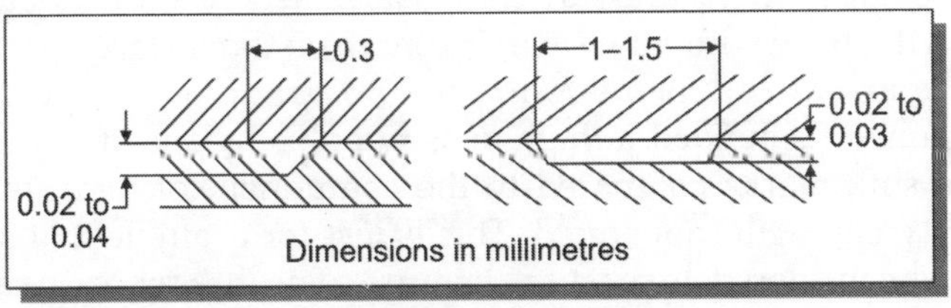

Fig. 2.64: Air vent

Sink Marks

Sink marks are the depressions on the surface of components. Sink marks generally occur where wall thickness is too much or there is a rib below the surface where sink marks occur. Sink mark may be reduced or removed all together by adjusting injection pressure, pressure stay, mold temperature. In case removal of sink marks is not possible by simple adjustment, then the product/component design would have to be reconsidered. May be it becomes necessary to do gas assisted molding. This may call for modification in mold design. Figure 2.65 shows few conditions of sink marks.

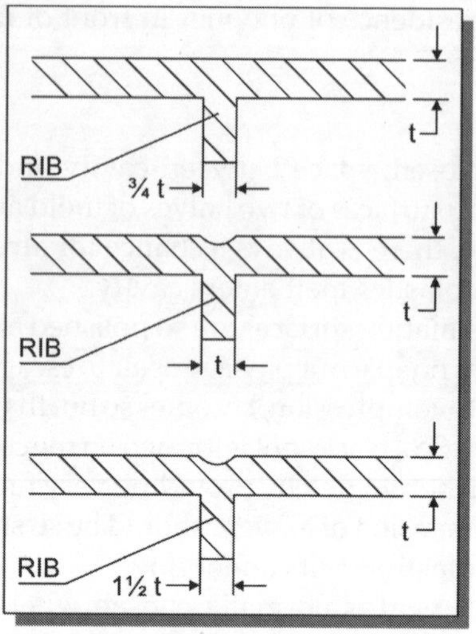

Fig. 2.65: Reasons of sink marks

Figure 2.66 shows gas introduction into the flow of plastics melt. Gas remains under pressure till plastics melt is solid enough to retain its shape without visible sink mark on the surface. Extent of sink mark also depends on type and grade of plastics. If a component is molded in high density polyethylene, it will have more sink marks compared to the component of high impact polystyrene with same mold. This is just for example that same mold is considered. In practice, there are some differences in cavity dimensions as percentage of shrinkage by volume is more for HDPE.

Flashes

Flashes are fine protrusions of plastics on molded components. Thickness of flashes may vary from 0.3 to 0.6 mm and even more. Length of protrusion may vary from 0.5 to 2 mm and even more. It is the plastics which flows in the gap between mating surfaces. Most likely spots of flash generation are meeting surfaces of two halves of mold, between ejector pins and their housing or even badly assembled fixed parts of cavity and cores.

Over heated plastics melt and high injection pressure is some times the cause of flashes. Figure 2.67 shows flashes in some plastics parts/components.

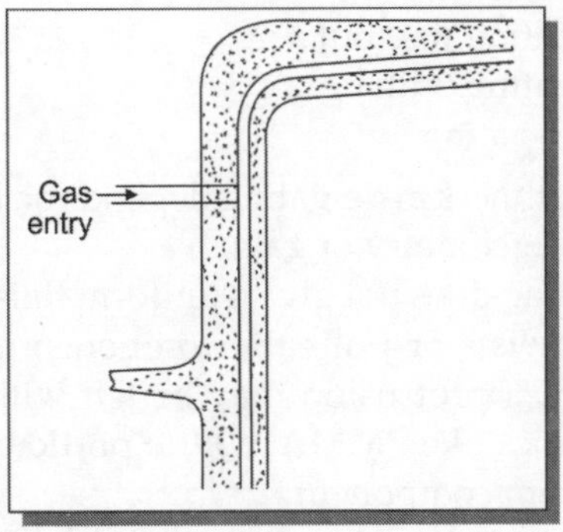

Fig. 2.66: Gas introduction in runner

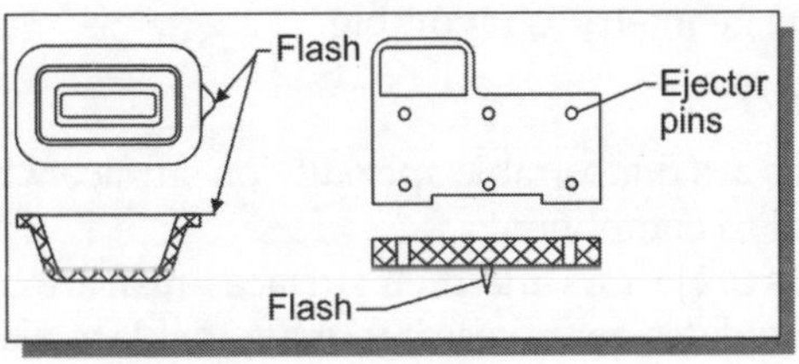

Fig. 2.67: Film/Sheet extrusion die

Flashes described so far are due to mold or material.

Another major cause of generation of flashes may be inaccuracy in parallelism of machine platen faces. If platens are not perfectly parallel, then some portions of meeting faces of mold would have more locking pressure as compared to other point where faces are meeting with lesser locking force. As a result, plastics flows easily for some distance between mating surfaces of mold halves.

Another cause of flash may be increased mold temperature or presence of some dust or dirt on the meeting surface of mold halves.

De-gating Not Smooth

Degating means separation of runner from component very close to component surface where plastics melt is being injected inside the cavity through gate.

Following methods are generally adopted for de-gating components:

- Manual de-gating without a jig
- Manual de-gating with jig
- Automatic de-gating

Basically, smoothness in de-gating depends on how fine the gate is and what is the geometry of gate.

Manual de-gating does not give a uniform finish. It depends on how a de-gater twists or pulls the component away. If plastics like LDPE is soft, a protrusion may be left with component. In case of hard plastics like PMMA, a little portion of material may be chipped off from component.

Components or articles where overall finish is of utmost importance then self-degating is generally adopted. It is achieved with submarine gates. In most cases, finish is good, if submarine gate is fine and its geometry is favorable.

Ejector Marks

Ejector marks are undesirable specially on surface which provide visual appeal to components.

Placements of ejectors and their surface finish are considered at the time of mold design. In case it is unavoidable to place an ejector pin at visual location then either the face of ejector pin is highly polished or some appropriate design is generated on the face of pin. Suppose cavity is to be finally spark eroded then it is done after putting the ejector pin in hole and be secured. It is placed in such a way that its face is also spark eroded together with cavity. In this way, ejector mark is camouflaged. Another method is to make a small design on the face of ejector pin in a symmetrical manner. Alternatively an alphabet may be engraved to give a visual feeling that it is purposely made at that particular position.

Whitish Marks

Whitish marks are actually micro air or gas bubbles which give a whitish appearance. It generally happens when plastics melt's residence time in plasticizing cylinder is more than desirable. Another reason may be that there are *dead* spots in either nozzle or plasticizing cylinder. Cleaning of these spots may solve whitish marks problem.

Clarity Below Standard

This deficiency is related to plastics materials like polymethyl methacrylate (PMMA), polycarbonate, etc. Many clear articles and components like lenses and covers are molded from these polymers. Light transmittance of PMMA is about 98%. If this value is less in molded component, then we say that clarity is below a maximum. It may be 92% or even below.

Micro-dots on the Surface

This deficiency is related to quality of surface of cavity. No matter how nicely polishing of cavity is done, micro-dots still remain. This is due to defective micro-structure of steel from which cavity is made. Solution to this problem is to replace cavity with a steel which does not have micro-pores in its structure. Special types and grades of steel are available which give excellent smooth polished surface without micro-pores, thus giving molded surface without micro-dots.

Distortion in Profiles

Distortion is unwanted variation in dimensions, straightness, angles, curvatures, etc.

Distortion mainly take place due to high temperature of melt, mold and premature ejection of components.

It also occurs if there is unusual resistance to ejection and ejectors pins or sleeve eject the component forcefully.

Safety Precautions

In injection molding shopfloor, one has to take care for safety of self, fellow worker, other persons present there and the machine and equipment. First things to do before handling a machine or equipment is to read its operating instruction manual. Wherever it is not clear then consult supervisor or technical officer.

Every machine and equipment manufacturer provides a checklist for checking, if safety gadgets are working fine.

In spite of best possible safety system, failures do occur, although rarely. So **never bring your hand** in between mold halves while injection molding machine is *ON*. If component is stuck and needs to be taken out manually then switch off the machine before bringing the hand between two halves of mold.

In case of low viscosity polymers at plasticizing temperature, like polyamide, there is a chance of splash during purging. So be careful, wear safety glasses and shield. Once the setting is normalized and there is no chance of splash, then safety mask may be removed.

EXTRUSION

Extrusion is a production process where polymer in the form of resins, pellets, granules and powder is fed to the hopper of machine known as extruder. This name is quite appropriate as machine thrusts out molten plastics. It is a continuous process as thermoplastics melt keep on coming out of an opening (II) without break as long as extruder is running. Basic working principle of an extruder is explained with the help of Fig. 2.68.

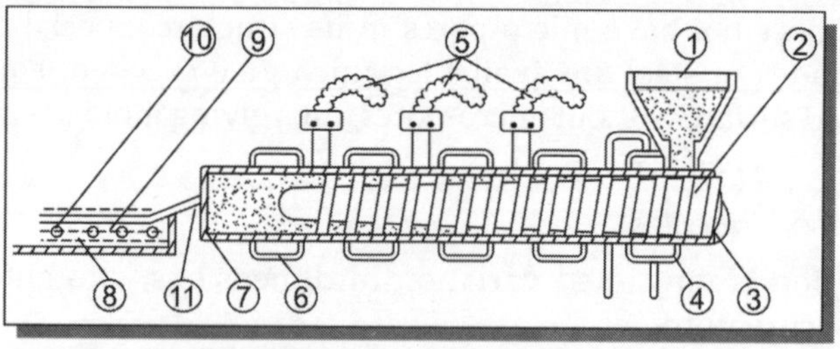

Fig. 2.68: Extruder section

Plastics granules are fed to extruder hopper (1) with the help of hopper loader. Granules directly drop down through a throat on to screw (3) which is placed inside a cylinder (2). Extruder cylinder diameter is fitted with a water circulating manifold (4). It is fitted just near the joint of hopper to throat. Along the length of cylinder, there are a number of band heaters (6) and temperature sensors (5). Initially, heater is switched *ON* to heat the cylinder up to desired process temperature. Once screw starts rotating, shearing action in melt and over coming friction generates heat to such an extent that heaters get shut down and cooling air starts circulating around barrel. (7) is the die which is fitted at the end of cylinder. Die is made with an opening of desired shape and dimensions. Typical sectional shapes of few extruded lengths are shown in Fig. 2.69.

There are innumerable possibilities of section designs. Dies are designed to give desired shape. Empiric formulae based on practical experiment are developed to determine most favorable dimensions of die opening so that after final cooling down, section and dimensions of product should be as desired. Sectional shape

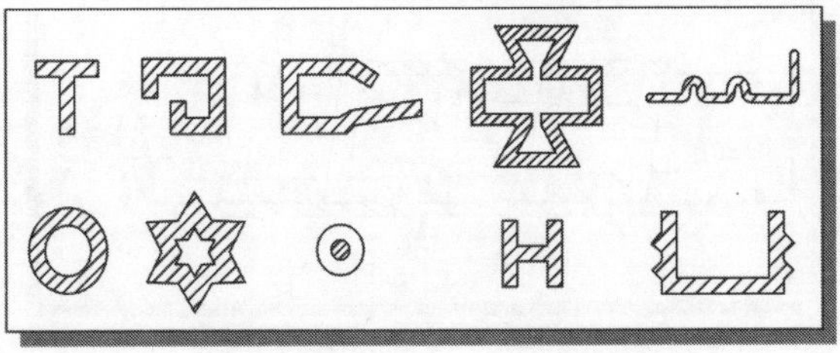

Fig. 2.69: Extruded sections (lengths)

and dimensions also depend on factors such as melt properties, temperature of melt, throw put, takeoff and cooling system. At the development stage of a particular product, all variables are recorded when desirable results are achieved with a particular die design. When ever production of a particular product is taken up, recorded parameters are set. Once extruder starts producing, fine tuning may be necessary to achieve best results.

Full length of the cylinder has a number of band heaters to heat the different temperature gradients. Temperature sensor provide signals to temperature controller to maintain particular value of set temperature. Temperature near the throat should not be to that extent that granules get softened and form a lump. This will block the passage for granules to drop down. For this reasons, a water circulating manifold (4) is provided to keep the throat area cool.

Function of an extruder screw is to convey granules forward, with available heat to gradually prepare a homogeneous mix continuously and force polymer mix out with sufficient force through filter screens and die. Length of extruder screw is made many times its diameter and is specified as L:D ratio (length and diameter ratio). L:D ratios may be 24:1, 32:1 36:1. An L:D ratio of 24:1 is commonly used.

Working principle of extruder screw (Fig. 2.70)

ϕ_1 = Bore of plasticizing cylinder

ϕ_2 = Diameter of screw

θ = Helix angle of thread

r.p.m. = Round per minute in clockwise direction when seeing from die end.

Depending on diameter of screw, clearance (ϕ_1–ϕ_2) is suitably given. It may range from 0.5 to 2 mm.

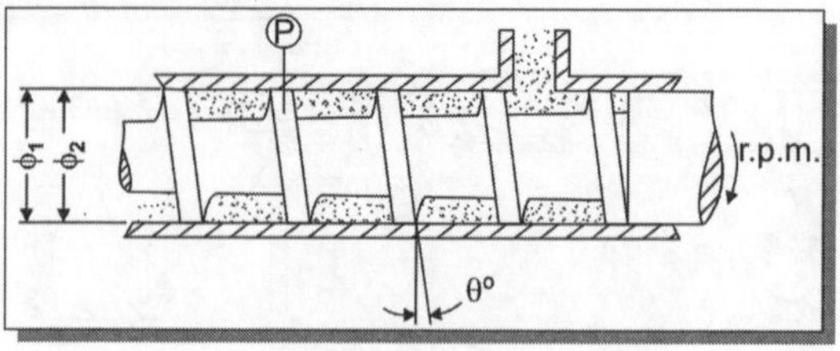

Fig. 2.70: Extruded screw

Let 'p' is a granule in contact with wall to thread of screw which is rotating in clockwise direction. There are two vector forces acting on granules. One is rotational and the other is axial. Combination of both the forces makes the granules churn as well as move forward toward die end. In this process, plastics granules, colorants, master batches, additives and modifiers get thoroughly mixed up to make a homogeneous melt till it reaches screens before the die.

Quality of melt also depends on length of plasticizing cylinder, angle of helix and r.p.m. of screw. Pressure developed in melt to pass through screw also depends on helix angle, screw's overall design and speed of screw in round per minute, which is generally 120 r.p.m.

Screw Design

Before a screw is designed, it is necessary to know what polymer is going to be extruded, its viscosity at melting temperature, coefficient of friction while in granular, semi-solid and melt forms. How sensitive polymer is to temperature variation of melt and residence time. It is also necessary to know as to how much throw put is required. Is a plastics prone to bleed gases and water vapors while it is being homogenized in barrel and forced towards filter screen.

Figure 2.71 shows a typical single flight screw. Flight length of screw, three zones of length are generally recognized as feed, transition or interchange and metering. End of feed zone towards drive system is just under throat to receive polymer granules. Function of feed length of screw is to push granules forward which are gradually heated. Softened granules enter transition zone where soft granules are gradually and thoroughly melted and homogenized. Melt now enters metering zone where it is further homogenized and brought to correct temperature and high

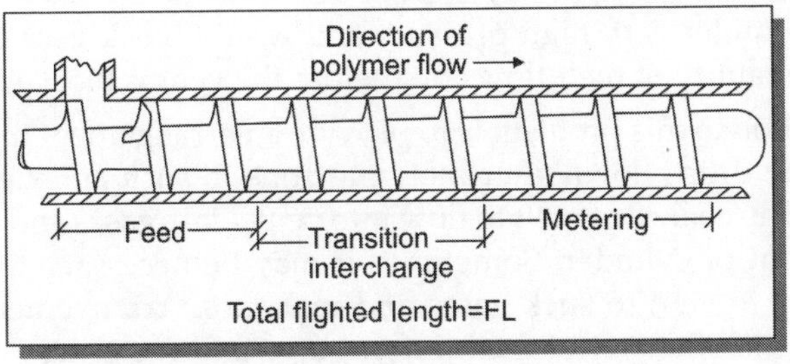

Fig. 2.71: Screw length functional sections

pressure to overcome resistance in flow due to filter screen and die. Function of screen is to arrest any contaminant which may be present in the melt. It also changes rotating tendency of melt to axial straight flow through die opening.

Figure 2.71 may also be represented by a diagram as shown in Fig. 2.72.

Extrusion of hygroscopic polymer such as polyamide (extrusion grade) may not give extrudate of good quality due to presence of moisture and gases in melt prior to coming out of die. It is because of the fact that during high pressure development in metering

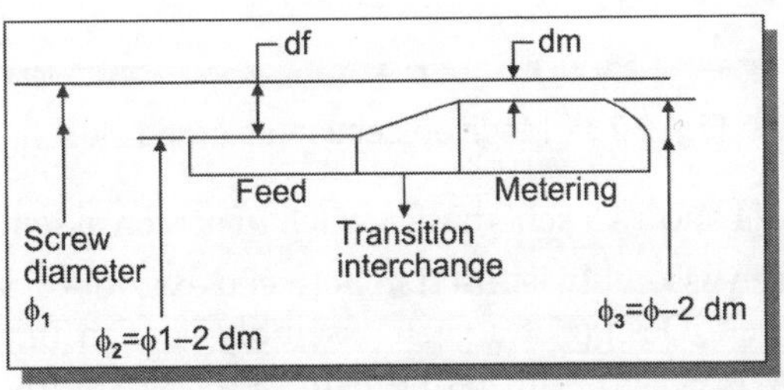

Fig. 2.72: Modified section

zone, gases and vapors evolve from melt and remain entrapped in the melt. This deficiency may be overcome by providing a means to remove gases and vapors before the melt inside plasticizing cylinder passes through filter screen.

A vent provided in cylinder after metering zone would not be of any use, rather would create problems. Melt in front of metering zone is under very high pressure and would block vent or even some quantity of melt flows out before the vent is blocked.

Solution to this problem is to provide an arrangement that melt pressure drops down after metering zone to such an extent that there is no chance of melt to flow into vent. Only gases and vapors bleed out of cylinder. Sometimes it may be necessary to apply vacuum to vent to suck gases and moisture. Arrangement is to provide another set of feed, interchange and metering zones in screw after first set. In this case, L:D ratio of screw may increase say, from 24:1 to 32:1. Idea of this type of screw construction is shown in Fig. 2.73

Once melt is made free of gases and water vapors, it again reaches metering zone of second set and attains desired pressure for melt to pass through screen and die to form extrudate of targeted sectional profile with specified dimensions, finish, strength and shape.

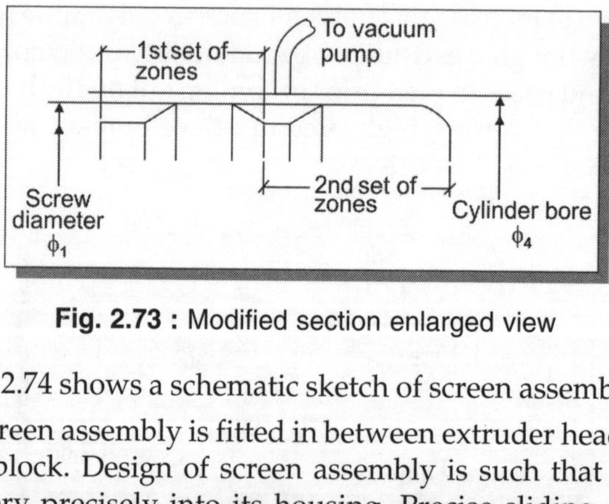

Fig. 2.73 : Modified section enlarged view

Figure 2.74 shows a schematic sketch of screen assembly.

This screen assembly is fitted in between extruder head and die holding block. Design of screen assembly is such that it can be slided very precisely into its housing. Precise sliding clearance does not allow melt to leak through clearance. All the melt passes through wire mesh and steel breaker plate. This plate of steel is strong enough to withstand high pressure of melt and supports wire mesh screen. Size of contaminants which can be arrested by mesh depends upon mesh size.

Another function of screen assembly is to create back pressure which is needed for uniform melting and proper mixing of polymer.

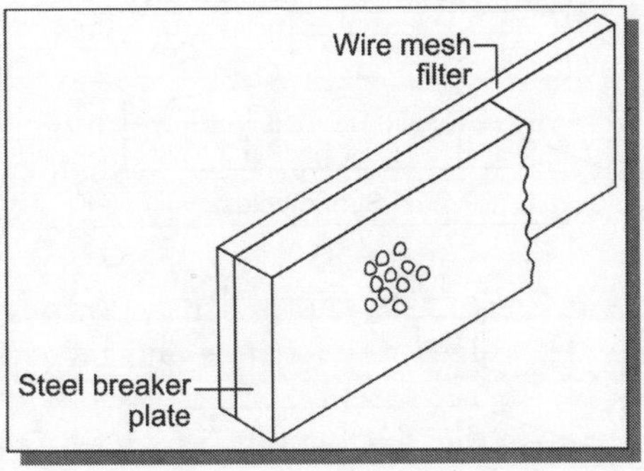

Fig. 2.74: Polymer filter plate

Extrusion process may be employed for conversion of plastics into:

- Compound
- Sheets
- Cast film
- Blown film
- Wire covering
- Tubing
- Pipes
- Co-extrusion

Compounding

In some injection moldings and extrusion shops, some grades and types of plastics in a number of colors and shades are used. Such processors usually procure plastics granules in natural color. These plastics granules are then extrusion colored in quantities of desired color and shade. Color pigments or master batches are mixed with natural color granules in a particular ratio before loading to extruder hopper. A special extrusion die (head) is used to get a number of colored strands (Fig. 2.75). Typically number of strands may be from 8 to 24. An experiment set up may have a 2 to 4 strand die fitted to a mini-extruder design of a typical die is shown in Fig. 2.76.

Die is made of two-halves (1) and (2). Centers of strand holes lie on meeting surfaces of two halves of the die. Die design may also be such that a separate plate may also be fixed between two halves

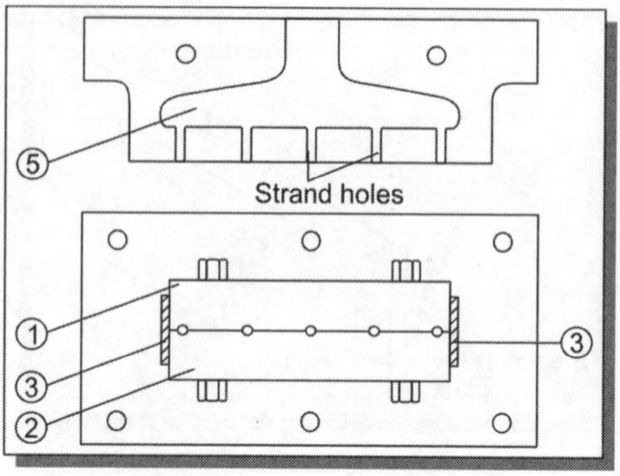

Fig. 2.75: Strands die design

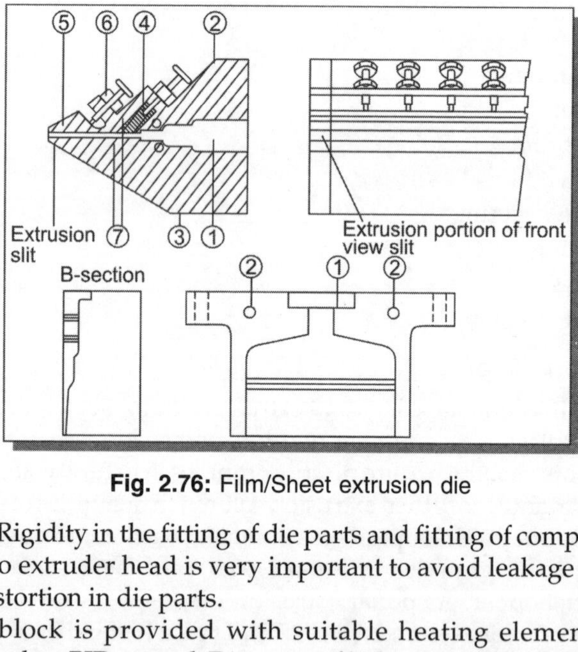

Fig. 2.76: Film/Sheet extrusion die

of die. Rigidity in the fitting of die parts and fitting of complete die block to extruder head is very important to avoid leakage of melt and distortion in die parts.

Die block is provided with suitable heating element with independent PID control. Diameter of holes typically may be 2 to 5 millimeters depending on the size of pallets or granules required.

Strands coming out of die are lowered into a water trough where cool water enters at one end and warm water goes out of trough at the other end. On cooling, diameter of strand get reduced say from 3.5 to 3 mm. At the end, strand passes through *drying oven* where water adhered to strand is removed by drying. After the dryer, strands reach granulators where strands are cut into small bits called granules. Compounding extruder process is also employed for mixing one or more polymers. Additives and modifiers may also be extruded to give polymer compound.

Sheets

Thickness of sheets is usually more than 0.25 mm. These are produced from a number of polymers such as polystyrene, high impact polystyrene, polypropylene, polyvinyl chloride (PVC), etc. Special extrusion die and take off auxiliary equipment are used to receive extrudate from extrusion die. Extrudate which is in the form of sheet and still soft, travels between combination of roller serve the purpose of giving the desired surface finish, thickness and cooling down the sheet during rolling process. Speed of rollers and their temperature is important as quality of finish and straightness of sheet also depends on this.

In practice, speed of rollers and output of extruder is synchronized by *fine tuning* of system parameters, such as extruder screw speed, melt temperature at entry to die, die lip temperature, die lip opening throughout its width, pressure by which melt is flowing inside the die.

Melt coming out of die lip swells immediately at exit. Sheet in soft state is taken over by *down stream* system. After cooling down, sheet thickness is stabilized. It is generally not necessary that value of sheet thickness is the same as opening of die lip throughout its width. This is due to swelling of polymer which takes place as extreme sheering action on melt while forced to pass through choking bar and die lip. Sudden release of sheer action at exit provides opportunity to polymer *chain* to attain its *normal length* causing it to thicken (swell).

Dies which are designed and fabricated for a particular type and grade of polymer may not give desired thickness, if polymer of other type is processed with the same die. It is, therefore, desirable that some *fine tuning* adjustments are provide in the die to make it more versatile.

In Fig. 2.76, (2) and (3) are two halves of sheet die which are bolted by nuts and bolts. Thickness and type of steels is so selected that there is no significant deflection due to presence of melt under high pressure. Choking bar (4) is provided throughout the width for adjusting the flow of melt to the die lip (5). By adjusting choking bar (4) in or out, pressure and speed of melt may be adjusted. Function of die lip (5) is to adjust the thickness of sheet throughout its width. Die lip is made of a steel having springiness so that slight adjustments (say 0.02 to 0.05 mm) may be made with the help of adjusting screws (6). There are a number of screws throughout the width to do adjustment according to measured thickness of extruded sheet. Once sheet of uniform thickness and surface quality is consistently achieved, all settings are recorded and fed to PLC (programmable logic control) for future repeat setting. Both upper and lower halves of die are provided with holes for fixing temperature sensors (7). Sheet may be produced with glossy or mat surfaces. Embossed surfaces can also produced by giving desired finish to rollers. Rollers are usually chromium plated for all types of surfaces finish. Purpose of chromium plating is to safeguard roller surface from rusting and to maintain shine. A sample of design impression on sheet is shown in Photo 2.10.

Films

Thickness of films is usually less than 0.25 mm. There are two methods to produce films of polymers such as LDPE, HDPE, PC, PET, PVC, etc.

Photo 2.10: Design impression on sheet

1. Straight
2. Annular

In straight method, extruded film comes of a die known as T-die. Basic shape of T-die is shown in Fig. 2.77.

(1) Is melt inlet of die
(2) Uniform cross-section of manifold
(3) Land length
(4) Holes for bolting two-halves of the die together

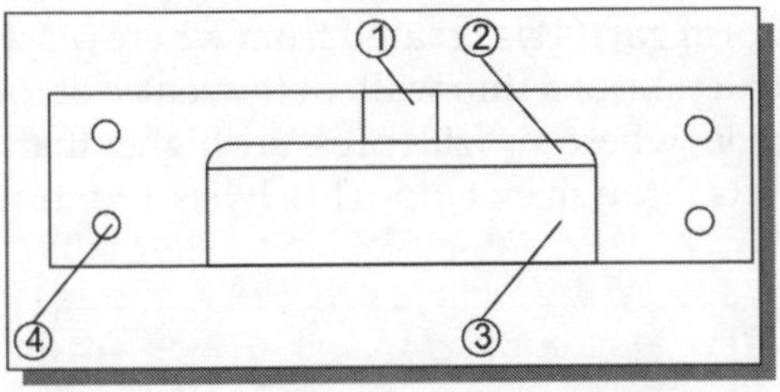

Fig. 2.77: T-shape extrusion die

Figure 2.78 shows basic design of another die which is called *coat hanger type*.

(2) In this case, manifold cross-section gets gradually reduced away from centre line
(3) Land length becomes shorter further from center line.

These types of dies are generally used for coating substrate like paper, aluminum foil, etc.

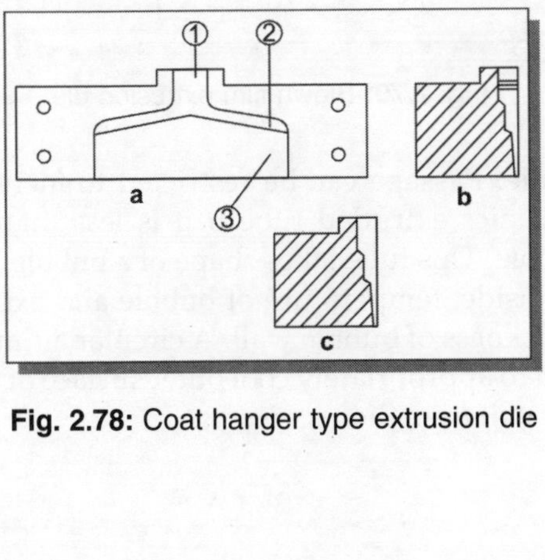

Fig. 2.78: Coat hanger type extrusion die

Down stream equipment trims the film to remove irregularities at the edges of film. Standard trimed width of film is maintained together with substrate.

Annular or Blown Films

Figure 2.79 shows a schematic arrangement of blow film extrusion die. (1) is extruder head where die is fitted. (2) is the outer body of die. (3) is a circular torpedo with three spider legs by which torpedo is precisely filled inside the die bore. Dimensions of die bore and torpedo's diameter are such that fine annular uniform gap (4) is created from where plastics melt exit in the form of a tube of a thin wall. In the centre of torpedo, there is a pipe (6) from where air with pressure higher than atmosphere enters and exits through exit pipe (13). It has a controlled opening

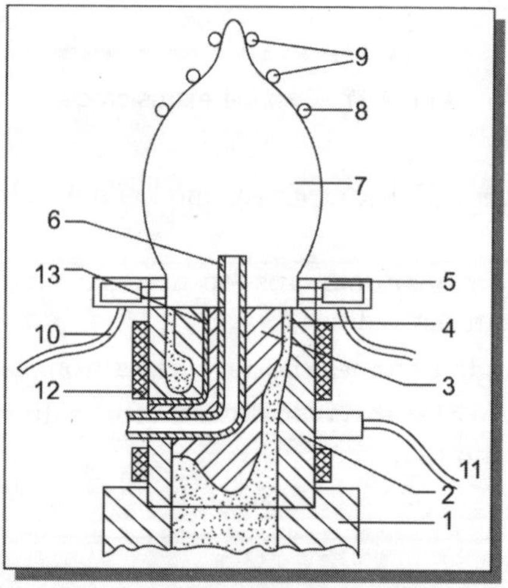

Fig. 2.79: Blown film extrusion die

outside the die. Passage can be restricted to increase pressure inside the plastics extruded tube if it is less than required to maintain bubble. This tube takes shape of a bubble. Appropriate air pressure inside, temperature of bubble and extrusion speed determine thickness of bubble wall. A circular air manifold (5) is also provided to appropriately cool outer surface of fine extruder tube just after exit from die.

After the bubble formation, film is taken up by hookup system. Film passes through roller or boards (8) which gradually make bubble film to become flat (Fig. 2.80).

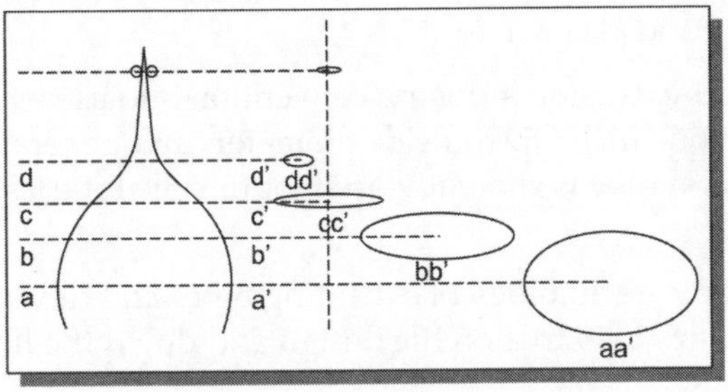

Fig. 2.80: Selection of blown film bubble

At ee' there are pinch rollers which give shape of *lay flat* film and press only so much that air above atmospheric pressure inside bubble does not escape out which is a means of constantly keeping the bubble formed. After pinch roller, *lay flat* film is brought down for auxiliary operating like corona treatment of surface, splitting *lay flat* into a flat film or any other operation (Fig. 2.81).

Pipe and Tube Dies

For watering garden, filling water drums or washing a car, flexible plastics pipe is normally used. It is not normally called plastic

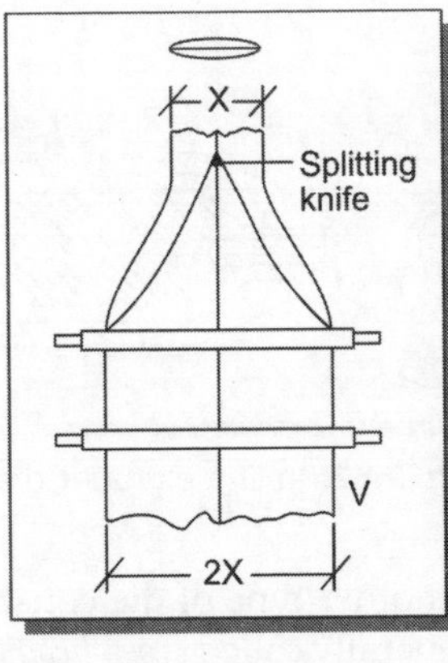

Fig. 2.81: Splitting of bubble

tube. Pathologists use a tube for extracting stomach juices for test. Note that this plastics tube is small in diameter (say 3 to 5 mm) and is *NOT* called pipe.

Terms pipe and tube are used interchangeably by various people, in industry and else where.

In plastics extrusion industry, convention is to use term pipe for more than one inch OD (outside diameter) and generally of rigid quality. Term tube is generally used for flexible tube less than one inch.

In principle, technology of extruding pipes and tubes is similar. Difference lies in extrusion die design and down the lines *take up* equipment.

Schematic sketch of a pipe or tube extrusion die is shown in Fig. 2.82. It is called *inline extrusion die*. Flange (1) is fitted to extrusion barrel head. (2) is a torpedo precisely fitted to bore of die. Its fitting is such that a concentric annular gap (3) is formed for extruded plastics tube or pipe to exit. (4) is an air passage to inject pressurized air (slightly more than atmospheric pressure). Purpose of this air is to avoid collapsing of tube after it exits the die opening (3). Pipe or tube then passes through cooling trough which has an inbuilt vacuum sizing fixtures. Purpose of this vacuum sizing fixture is to maintain uniform OD without oval shape and variation in size.

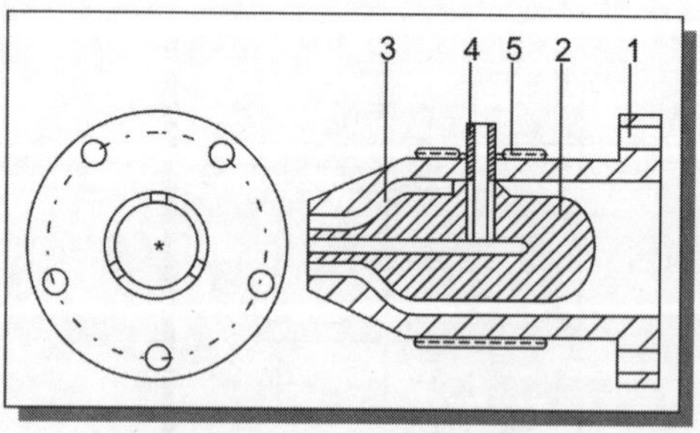

Fig. 2.82: In line extruder die

A basic feature of another type of die construction is shown in Fig. 2.83. This die is generally called *cross-head* die, because plastics melt enters at right angle to the axis of die.

(1) is flang to fit extruder head. (2) is cone or torpedo, (3) is circular and concentric opening or exit of extrusion die and (4) is the hole for injecting air for the purpose of not allowing pipe or tube to collapse. Die ring (5) is radialy adjustable to achieve uniform wall thickness of pipe or tube.

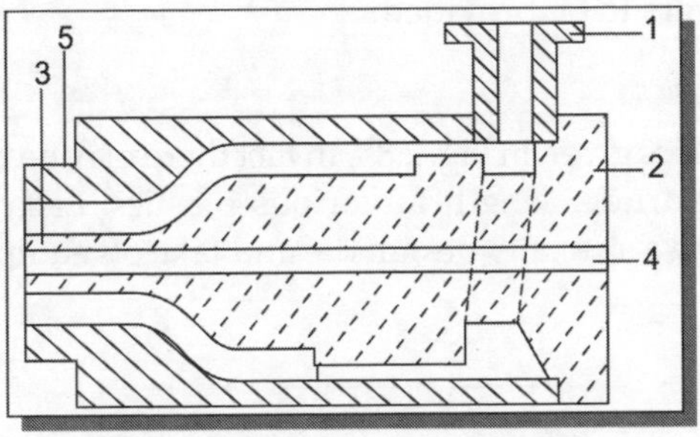

Fig. 2.83: Right angle extruder die

This type of die can also be used for coating wire with plastics. Die for this purpose is modified to suit the purpose. Desired modification is shown in Fig. 2.84.

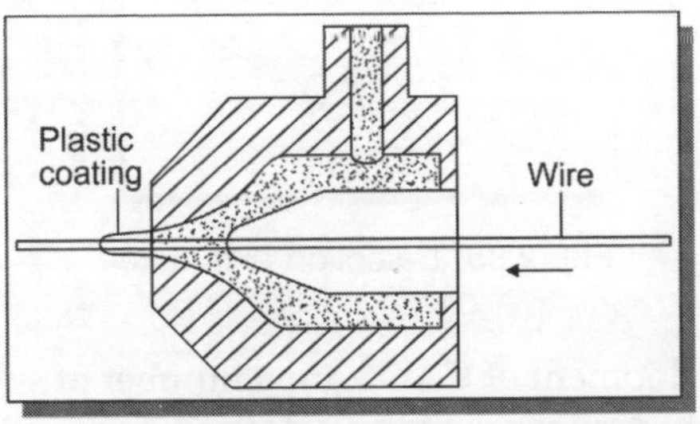

Fig. 2.84: Wire coating extruder die

This modification enables plastics melt covering to adhere properly to bare metal wire surface.

Multilayer Sheet and Film Extrusion

Sheet having more than one layer is a multilayer sheet. Multi-layer sheets or films have a combination of advantages. A sheet of ABS (acrylonitrile butadine styrene) having a layer of PMMA has impact strength and gloss of ABS and PMMA layer provides protections against UV (ultraviolet) radiations. In case of blown film, a layer may be provided to resist UV radiation.

Multi-layer sheets or blown films are provided by a set up where a number of extruders are connected to specially designed and made die. Numbers of extruders depend upon the numbers of different layers to be provided.

Profile Extrusion

Idea of profile is given in Fig. 2.85. In fact there may be innumerable profiles of extruder length in various plastics. Each profile and polymer need a specially design die and take off equipment down stream.

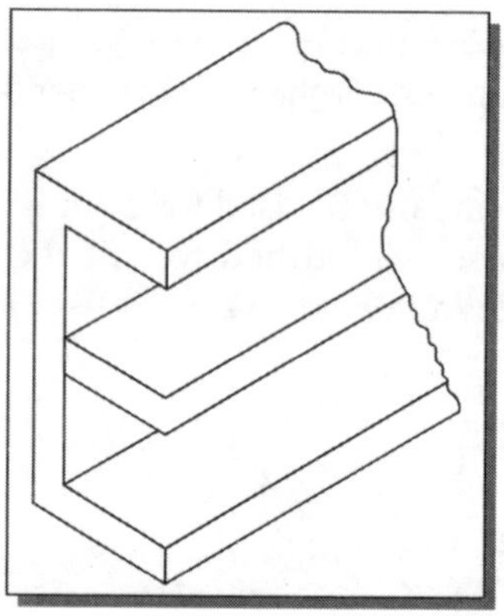

Fig. 2.85: E-section extrudate

Prior to development of CAD with a number of software tools, it was difficult to design a die to give desired dimension of finished product. Die exit dimensions were finalized by trial and error method. In between trials, costly modifications were carried out till correct dimensions of sectioned profile were obtained.

Resent developments in CAD and other software tools, it has become possible to design and make a profile die to give product of specified quality and dimension.

Dimensional accuracy and finish of extruded profile are influenced by following factors.

- Properties of plastics being extruded. Predominant properties are melt flow index, specific gravity, heat transfer rate, etc.
- Shearing action in the flow of polymer at various exit portions in die.
- Places where shearing action is severe, swelling of that portion at exit would be more .
- Non-uniform section of profile tend to develop stresses due to uneven cooling and shrinkage. This is minimized by effective design.
- Development which is quite a complicated exercise.
- Use of cooling arrangement in the core of die.
- Provisions of sizing jig in downstream take off equipment.
- Dimensions of sizing unit are very near to finished dimensions of product. It remains under cooling water in trough near the exit of extrusion die in such a way that extrudate is still in pliable state so that it gets hard enough while passing through vacuum aided sizing jig, dimensions of cross-section are stabilized.

THERMOFORMING

This process is used to produce shallow or deep articles such as glasses, serving trays, inner linings of refrigerators, large panels and a variety of packing items from sheet of materials such as PP, PS, PVC, PET, PC, ABS, SAN, etc.

Nomenclature *thermoforming* implies that application of heat is one of the main requisite for carrying out thermoforming operation.

Figure 2.86 shows three thermoformed articles.

In Fig. 2.86, a typical plastics glass is also shown, which is generally used for serving tea or coffee by vendors at public places such as railway stations, bus stands and roadside tea stalls. Its production by thermoforming operation is explained with the help of Fig. 2.87.

(1) is a wide clear PP or PET sheet coming from roll stock. It is intermittently pulled by draw system which indexes with a predetermined *stay* time. In this typical example, sheet is 0.25 mm thick and 245 mm wide. Complete thermoforming set up consists of following stations.

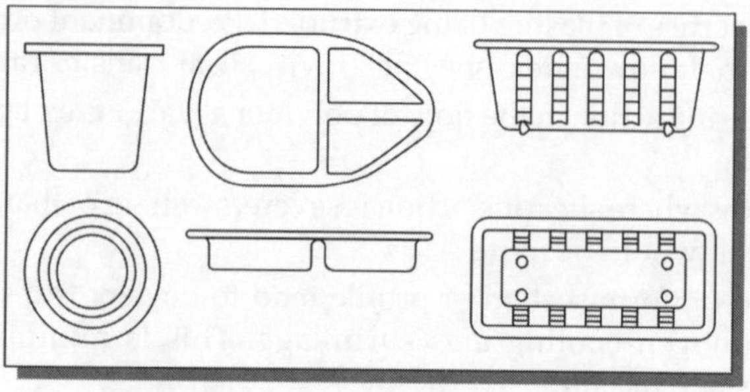

Fig. 2.86: Thermoformed components

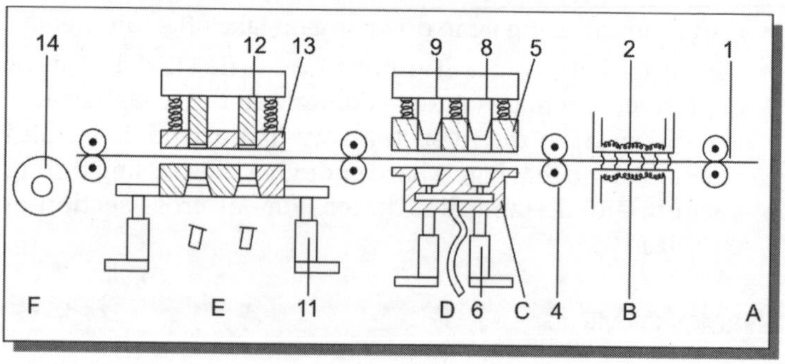

Fig. 2.87: Thermoforming system

A – Roll feed stock with tensions
B – Heating station, heating both side simultaneously
C – Forming system
D – Frame support roller and guide channels
E – Trimming station, glasses to fell down
F – Take off system
G – Chopping and granulation system (not shown)
H – Mixing with virgin plastics powder and granules for sheet extruder (not shown)

From economy considerations, thermoforming plants are designed and made to do fast production of bulk quantities. Width of sheet is so chosen that 5 to 7 or even more glass impressions are possible. In a typical example, five impressions of 55 mm diameter glass would need a sheet/roll width of about 371 mm. Across the width, there are two rows of impressions. One row has 5 impressions. Four impressions in another row, allowing closer placement so that percentage of scrap generations is as less as possible. Distance in between central lines (P) of impressions is 60 mm. So pitch of indexing of sheet would have to be 120 mm (Fig. 2.88).

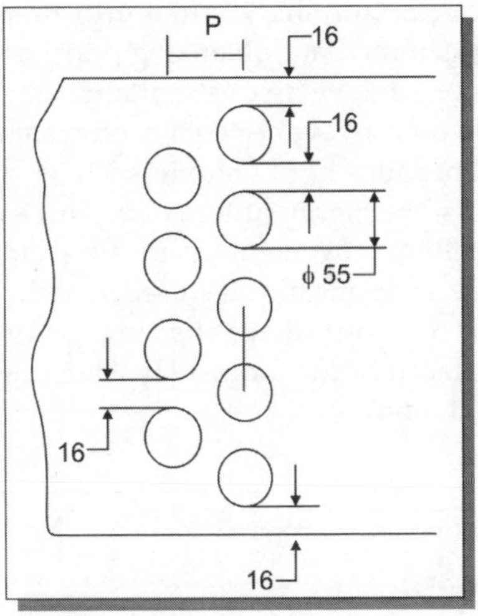

Fig. 2.88: Blank impression layout

Referring to Fig. 2.90, a number of heating zones are there, having selected elements such as cal rod, quartz, ceramic, infrared and nichrome.

Some are above the sheet surface and other below the sheet. Length of heating panels and distance from sheet are so kept that gradual heating of sheet is possible so that sheet is softened up to the core. Precise temperature sensors and controller are set to maintain proper temperature of sheet so that when heated portion of sheet reaches station D its softness is in pliable condition for thermoforming. As soon as heated portion of plastics sheet reaches station D, upper tool with pressure pad in advance condition presses the sheet. Immediately after pressing vacuum switch operates to create vacuum in cavities. At the same time, draw punches draw the sheet down to about three-fourths depth while vacuum is sucking the sheet to take shape of die. So both the actions thermoform the glasses. Before the beginning of forming operation, indexing of sheet stops. As soon as thermoforming gets completed, sheet with formed glasses is lifted by guide channel lifters so that formed glasses already attached to sheet are clear of obstruction and then gets indexed. Alternatively, forming station may also move up and down.

On E station, guide channel gets lowered and formed glasses already integral part of sheet enter into blanking ring dies concentrically and immediately blanking punches come down and blank out brims of all the glasses simultaneously. Blanked down glasses fell down on conveyor or collection crates. Modern plants are designed to produce high volume such as 50000 glasses per hour. Such plants are highly automated with sensitive controls over heating, cooling, movements, punching, etc.

Figure 2.89 shows packing tray. It is used to pack electronic gadget. An information cut out sheet, having printed matter and photographs is placed inside for display. Hanging holes are made by piercing tool station.

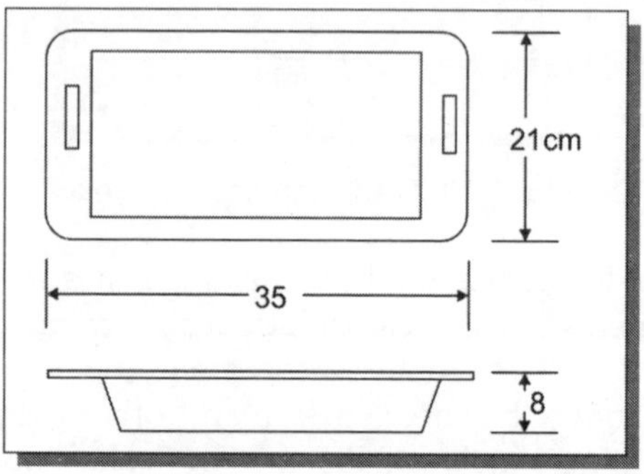

Fig. 2.89: Big tray, forming and draw

Dimensions of typical tray are provided to enable the reader to have an idea that for such dimensions only vacuum forming is sufficient. No draw punches are needed in this simplifies mass production set up. Figure of this item is also provided. It can also be produced by a single thermoforming station. Photo 2.11 shows such a machine.

Cut pieces of sheet may be manually fed to the machine. A frame presses the sheet over the vacuum die brim to make it vacuum tight over the brim surface. After proper heating of sheet, vacuum

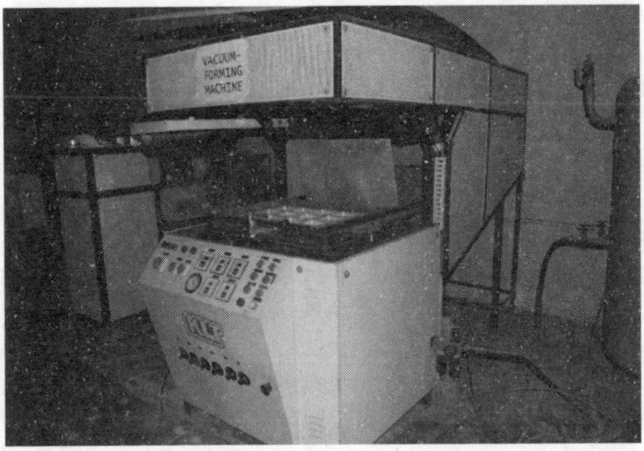

Photo 2.11: Vacuum forming machine (*Courtesy*–University Polytechnic AMU)

valve is manually operated which connects the die to a vacuum chamber of suitable capacity. Since the chamber is already evacuated and has a high vacuum, it sucks the air from the die very fast. Consequently, PET sheet in pliable state takes the shape of cavity. It is then allowed to cool down for few seconds to maintain the shape permanently.

A typical refrigerator door lining may be produced from HIP (high impact polystyrene) or ABS sheet. Thickness of sheet may vary from 2 to 3.5 mm. It may be noticed that lining has deep and narrow hollow ribs, which are difficult to thermoforming. It needs application of vacuum, mechanical pushing and high pressure air to attain details of contours of ribs.

Basic construction of such a set-up is shown with the help of Figs. 2.90 and 2.90a.

Figure 2.90 shows schematic arrangement of a sheet heating furnace. It consists of two sections A and B. Section A has low

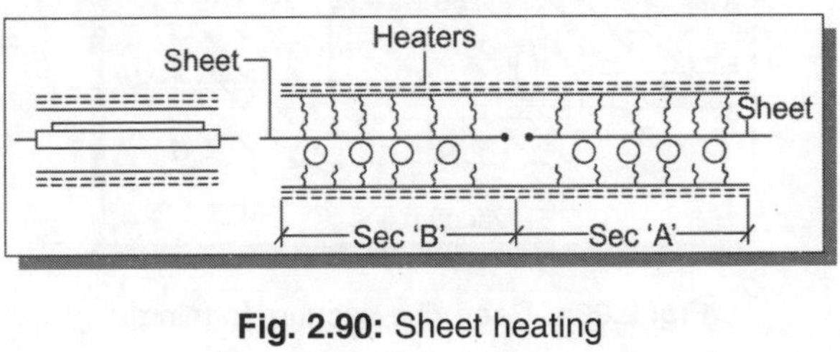

Fig. 2.90: Sheet heating

heat radiation to pre-heat the sheet for final heating in section B. Sheet is automatically transferred from section A to B. Temperature of sheet in section B is so attained that it becomes pliable. After a per-set time, sheet is then transferred automatically on thermoforming die.

In Fig. 2.90a, a portion of die or mold is shown. (1) is lower portion of die on which pliable sheets arrive. It has cavities (8) of suitable depth, width and taper to give proper shape and dimensions to ribs as per drawing. These cavities have small holes (5) to connect them to vacuum chamber (6). All the ribs, cavities small or big have similar holes. Diameter of hole is kept as big as possible to get maximum sucking action. At the same time, hole diameter should not be so big to leave a mark on product. Since the sheet is thick and the rib is too deep and narrow, only vacuum draw would not give desired result. It is, therefore, necessary to use plugs/punches (3) to push the sheet inside the cavity (8). In addition to vacuum inside the cavities, high pressure air is also injected through a hole in draw punches (3). Air pressure is applied to draw of sheet inside the cavity, when it is about to be completed. After thermoforming is completed and formed sheet is completely rigid, it is then taken out of cavity block (1) and automatically transferred to trimming station where excess sheet edges are trimmed out to give final product, in this case lining of refrigerator door.

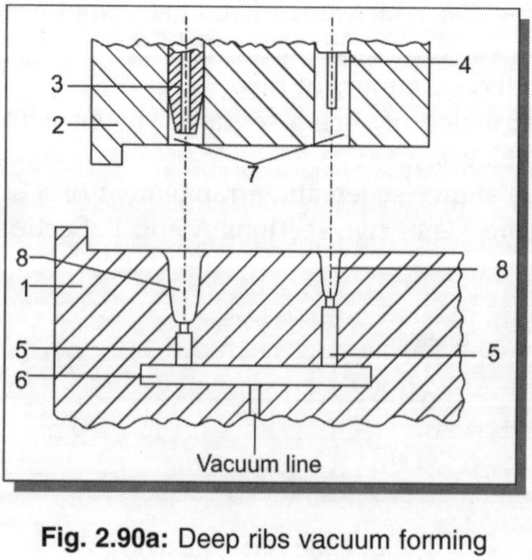

Fig. 2.90a: Deep ribs vacuum forming

Sheet Draw Behavior

Figure 2.91 shows, pliable portion of sheet is a circular disc of ϕ mm diameter. Weight of disc would be

$$W = \text{Vol} \times \text{density}$$

$$= \pi/4 \times \phi^2 t l$$

Let $\phi = 10$ cm

$t = 0.05$ cm

$\ell = 0.92$ gm per cubic centimetre

Hence $= 3.14 \times 0.25 \times 100 \times 0.92$

$= 72.2$ gm

Thermoformed cone shown in Fig. 2.91 has diameter ϕ. So the weight of the cone would also be 72.2 grams, irrespective of the shape or contour of cone.

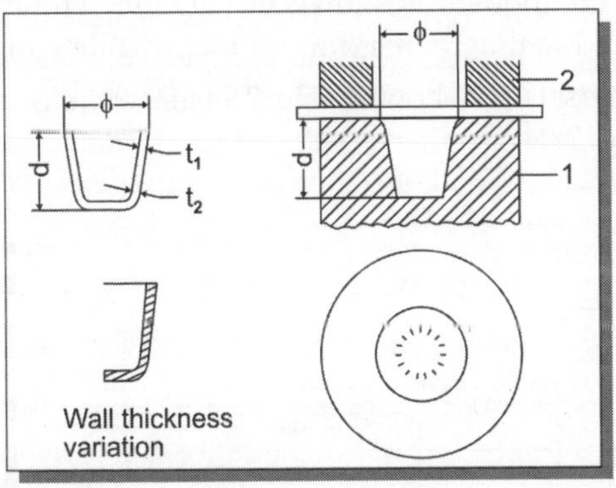

Fig. 2.91: Wall thickness variation in vacuum forming

During thermoforming operation by applying vacuum, area of *disc* ϕ increases while weight remains the same. Consequently, wall thickness of cone gets reduced. Reduction in wall thickness is not uniform. As soon as hot sheet in pliable condition come in contact with die, brim ϕ of cone looses heat quickly. Consequently, during stretching thickness of sheet gets reduced more at t_2 as compared to t_1. t_2 may become so thin that cone wall becomes soft and become difficult to hold.

In case it is needed that t_2 is thicker than t_1 then other method of thermoforming may be employed. Its basic arrangement is shown in Fig. 2.92.

After thermoforming, cone with extra drawn portion would be taken out from die. Cone would have to be trimmed out at ϕ from extra drawn portion. Depending upon shape and size of forming die, design is improvised.

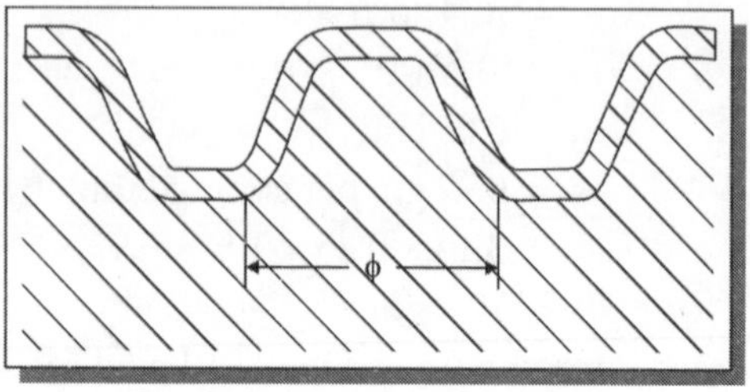

Fig. 2.92: Reverse vacuum forming

A typical component is shown in Fig. 2.93. How should it be thermoformed so that there is minimum possible scrap?

A schematic set up is shown in Fig. 2.94 to thermoform component shown in Fig. 2.94.

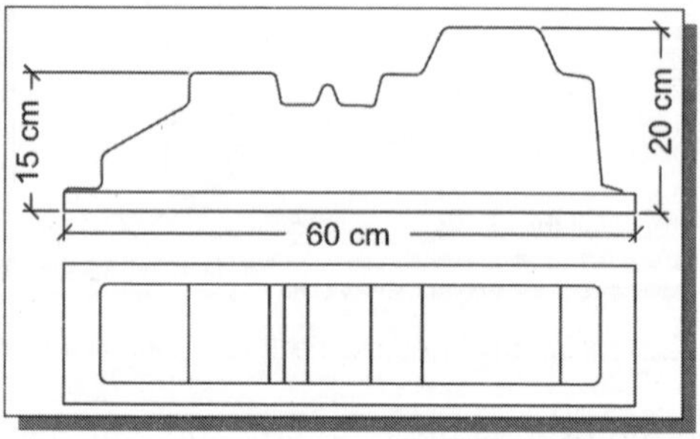

Fig. 2.93: Vacuum formed large component

Sequence of operation for thermoforming of component as shown in Fig. 2.93 would be as below (*see* the set up in Fig. 2.94).

- Sheet duly heated is brought over die (1).
- Clamp (2) presses the sheet over die surface only so much that air or vacuum does not leak.
- Air with controlled pressure and volume is injected through port P_1.

- Due to air pressure, sheet in pliable condition takes the shape of a *dome*.
- At the same time, plug gets lifted very near to sheet.
- Simultaneously, port P_2 gets connected to high vacuum chamber with high capacity so that hot polymer sheet gets wrapped over the plug very fast.
- Figure 2.94c shows formed component still an integral part of sheet.
- Clamps (2) are removed and vacuum is cut and plug goes down.
- Sheet is shifted to trimming die where component is trimmed out of sheet.

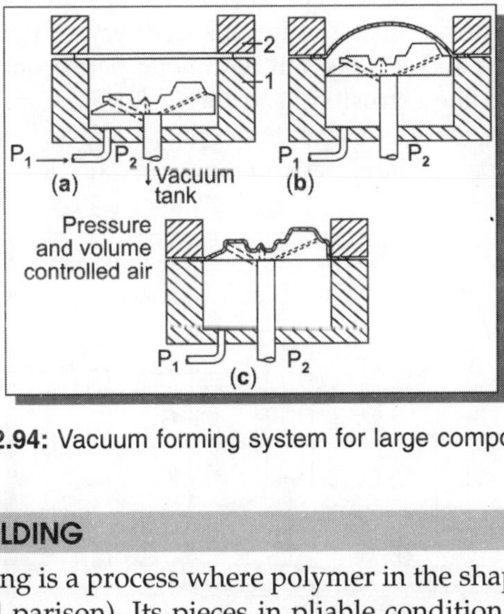

Fig. 2.94: Vacuum forming system for large components

BLOW MOLDING

Blow molding is a process where polymer in the shape of tube or pipe (called parison). Its pieces in pliable condition are trapped between two-halves of cavity blocks. High pressure air is injected in the centre of parison. Since hot parison is in pliable condition, it expands in all the directions only to stop at the walls of cavity. Parison after expansion assumes shape and size of cavity. On touching cavity walls, expanding parison gets cooled to achieve permanent shape of a hollow article.

A large number of varieties of articles like cold drink, mineral water and beverages, bottles, small and big jerry cans, cosmetics, medical, agriculture and technical items are produced by a blow molding process.

Blow molding is carried out by different methodologies, which are as follows:

- Continuous extrusion blow molding
- Intermittent extrusion blow molding
- Injection-blow molding
- Preformed stretch blow molding

Figure 2.95 shows a typical continuous extrusion blow molding system. Extruder (1) continuously feeds melted plastics to extrusion head (2). This head delivers melted plastics in vertical downwards direction in the form of a tube (3), which is called parison. Diameter, wall thickness of parison depends upon shape and adjustment of extrusion head, vertical continuous delivery of parison with a predetermined and set speed in centimeter per second.

Underneath the extrusion head there is a blow mold handling unit (5) which also carries parison cutting knife and high pressure air blowing nozzle. Handling unit moves up and down, mold (4).

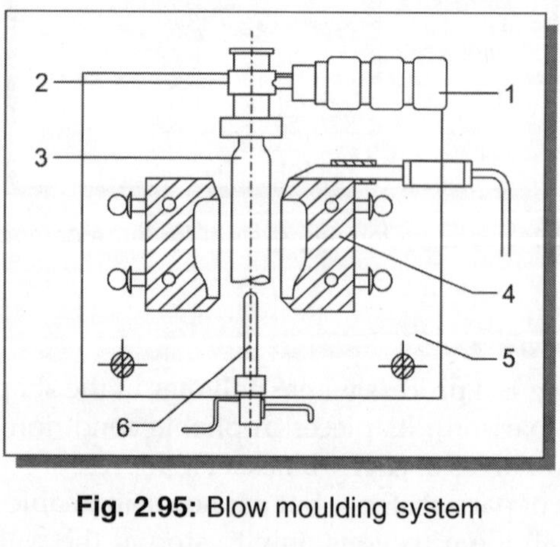

Fig. 2.95: Blow moulding system

Closes and opens, high pressure air nozzle/pipe (6) also moves up and down in relation with mold. Cutting knife also operates to cut parison. During continuous blow molding process, the following parameters are accurately controlled by sensitive sensors and close loop circuits.

- Parison diameter
- Parison wall thickness
- Parison temperature
- Parison downward speed

Sequence of operation is as under:

- Mold carrier is in up position.
- Mold is opened enough to allow parison to move down between the two halves of mold.
- Air nozzle is in up position in relation with mold.
- Parison moving down with a specified and set speed.
- After a specified length of parison comes out of extrusion head, mold system starts moving down with parison speed.
- Simultaneously mold closes.
- Simultaneously parison is cut.
- Mould gets shifted to one side, enough to make room for continuous parison extrusion.
- Mould opens and blown article is ejected together with scrap (flash) inherent to process.
- Immediately after ejection, open mold moves back to take next length of parison.
- Process then repeats itself.

Materials like LDPE, LLDPE, HDPE, PP, HM HDPE, can be extrusion blow molded. Extrusion blow molding process is also suitable for PVC. PVC is very sensitive to heat. It can degrade rapidly, if gets overheated.

Intermittent Extrusion Blow Molding

This technique is generally used for large containers like 25 liters jerry can. Figure 2.96 shows a typical arrangement of intermittent (non-continuous) extrusion blow molding.

- In closed condition, axis of mold cavity (7) is coaxial with axis of parison extrusion die (5).
- There is no vertical movement in mold. It only opens and closes.
- There is a hollow vertical pin (8). Its axis coincides with axis of mold cavity *mouth* opening (threaded opening) in closed condition. Its function is to inject high pressure air or nitrogen (as the case may be).
- Mould closes completely with parison inside.
- Extruder (1) is continuously feeding melted plastics in accumulator chamber 2.

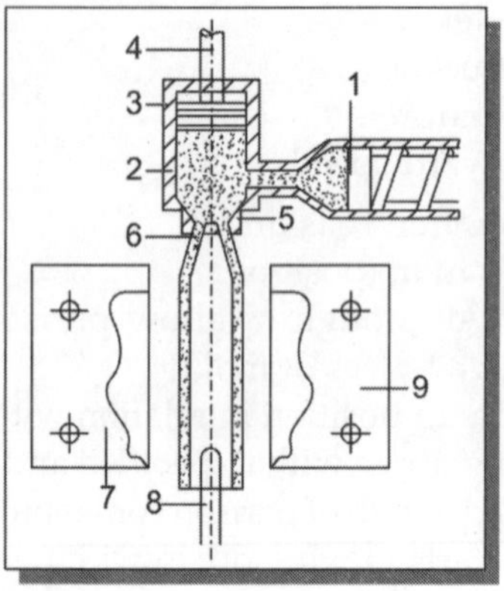

Fig. 2.96: Parison between mould halves

- Plunger (3) (operated by rod 4) inside chamber pushes melt down with a specified and set speed through extrusion die to form a parison. It is then blown.
- Parison on the top is cut by knife.
- Moulded item cools down to ejection temperature.
- Mould opens and blow molded item is ejected.
- Next parison immediately arrives between the open halves of mold.
- Process then repeats itself.

Items come out with flashes which can be removed manually or by automation. In case of automation, blow molded article with flashes is automatically picked up from ejection point and transferred to another station where de-flashing takes place. In case of extrusion blow mouding process, it is possible to mold articles with handle formation. Limitation with this process is that neck (opening) dimension and shape cannot be maintained with precise tolerances. Sophisticated blow molding items for technical purpose are generally required to have uniform wall thickness with close tolerances. Sometimes shape of article is such that blowing of parison varies greatly. Wall thickness remains thin where blowing is more and thick where blowing is less. To overcome this deficiency, to a great extent, sophisticated machines are used where variable thickness of parison wall can be extruded. It is controlled by microprocessor. Predetermined *profile* of wall thickness is set in computerized control panel which activates extruder die mechanism to generate different wall thicknesses at different portions of full length parison.

Injection Blow Molding

As the name suggests, first melted plastics is injected into a cavity to form a molding with precision neck having threads or locking design. Remaining body of injection molded component still kept hot on the core is transferred to another blow molding station.

As soon as mold closes, high pressure air from core passage enters to blow the injection molded part to take shape of blow molding cavity. Blow mold opens and core system indexes to reach third station where ejection of blow molded article takes place. Figure 2.97 shows basic system of injection blow molding machine.

Indexing turret carries three similar cores of suitable design and dimensions. There is arrangement to sense and maintain predetermined temperature. Also, all the three cores have air valve system to inject high pressure air. There are following stations.

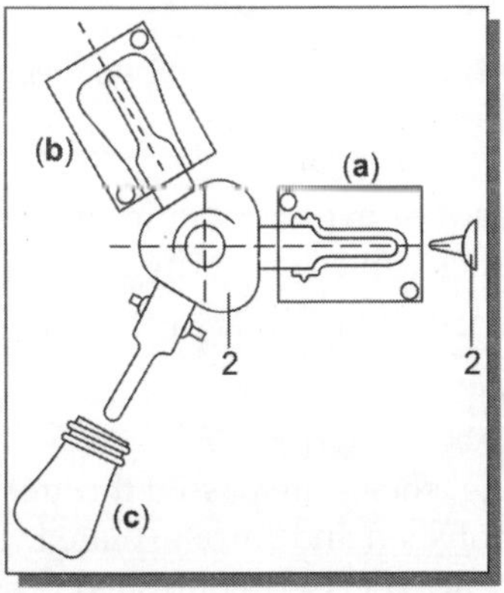

Fig. 2.97: Injection blow moulding

A. Injection of plastics melt where molded thread or locking portion is quickly cooled down. But portion to be blown on next station is kept at blowing temperature.
B. Mould opens and turret gets indexed.

At this station, a blow mold closes over the already injection molded perform. High pressure air or nitrogen is injected through core to blow the preform to take shape of cavity. Blown item is cooled down in cool cavity. Mould opens to provide enough space for blown item to move out of blow mold, when turret gets indexed to station C where blow molded item is ejected. So each indexing gives a component. Injection blow molding can also be done in another way. Preforms are first molded and collected. They are then heated to stretch blowing temperature and fed to stretch blowing unit. In this unit, perform is stretched by means of a rod and then blown. This process is specially used for producing PET bottles. Stretch blow molding improves mechanical strength of bottles, improves surface finish and gas permeability. With injection blow molding process, handled containers cannot be produces.

As said earlier, precision thread or locking system in product is produced with precision.

CALENDARING

Calendaring is a process through which following are achieved:

- Plastics sheets of uniform thickness with close tolerance of 0.03 mm.
- To impart smoothness and shine to one or both the surfaces.
- To give a variety of texture to plastics films and sheets
- To produce multilayer sheets
- To laminate surface of paper or fabrics by thin plastics film
- To laminate plastics woven sheets by plastics film on both sides

Basically, calendaring process is carried out in two ways:

1. Hot calendaring
2. Melt calendaring

In hot calendaring, sheets are passed through a number of rolls which are highly polished and chrome plated. Their temperature, speed and pressure are precisely controlled.

Across the width of sheet, sensitive and reliable sheet thickness sensors are installed to servo-control the sheet thickness. Gap between the rollers is precisely adjusted by means of automatic system which is controlled by servo-sensor system for thickness measurements and control (Fig. 2.98).

Calendar machines are usually used in conjunction with other machines such as plastics dispensing unit, set of rolls which have either highly polished mirror bright surface, texture or mat chromium plates surface. Photo 2.10 is a typical example of

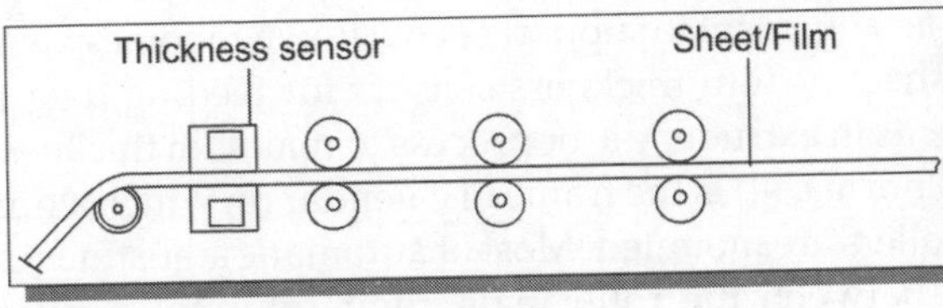

Fig. 2.98: Sheet thickness sensor

calendaring PVC flexible sheet with a textured design. This is the example of sheet which is generally used as table covers, curtain, etc. Another example of calendaring is covering of one surface by lamination process. It can be seen from Photo 2.12 below.

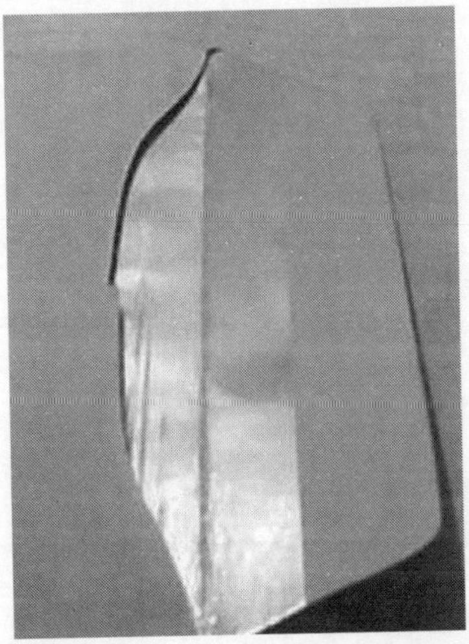

Photo 2.12: Lamination on paper

It is a paper envelop which has a very fine layer of polymer laminated on one side of paper sheet from which these envelop are made. Such type of plastics coated/laminated sheets are used for so many other purposes.

These polymers are generally used to produce rigid plastics films. Soft films are also calendared out of PVC, PE, PEV, etc.

Figure 2.99 shows basic arrangement of calendaring plant. (1) is an extruder with sheet or film producing die head (2). Sheet or film in between die exit and entry to calendar (4) rolls pair (5), there are arrangement to support or cool down to some extent. There may be sheet or film thickness sensors for feeding data to auto-adjustments in extruder die or process to maintain thickness under specified norms. (4) is the frame of calendar on which a number of pairs of rollers are mounted. Most of automatic adjustment systems for gaps betw een the roller pair, their rotating speed, surface temperature, etc. are automatically controlled. Pair of roller (8) or (9) are exit roller by which final thickness and surface finish are established. (10) is automatic remote controlled thickness controlling system. Thickness is continually displayed on an array of indicator and control buttons and switches.

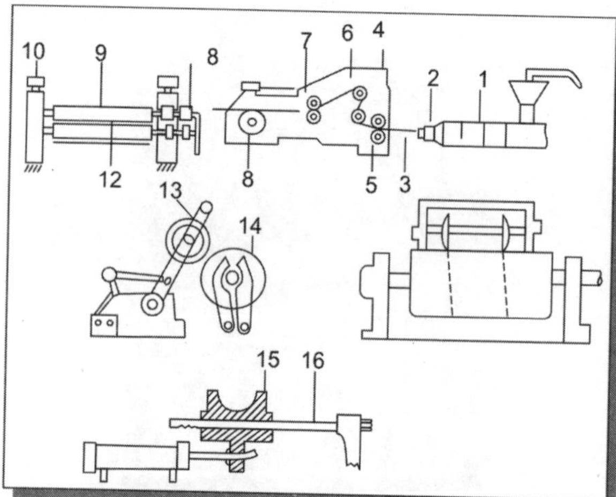

Fig. 2.99: Calendaring system

Although thickness of sheet or film is servo-controlled. Still there is a facility for supervisor to intervene and to carry out some adjustments. This is done by pressing some buttons while watching various meters on control panel.

Sheets and films of full width are wound (roll)on winder (14). After a preset diameter of wound is reached, sheet or film is automatically cut. Winder (14) moves away and placed over the stand (15) with the support rod extraction. Small width rolls are now manually or automatically packed in wrappers. In its place, another winder reaches to take over winding afresh. First winder reach under a slitting system where full width of sheet/film is slitted in two or three portions as per need. Big plastics woven sheets are also laminated on both sides by thin film of LDPE or LLDPE. It makes a good sheet for covering bags of grains in warehouses and general purpose covering.

Terpoline sheets are produce by covering cloth with melted plastics layers and then passing through calendar rollers where terpoline is pressed, smoothened and thickness is made uniform.

Hot calendaring is carried out by drop spreading syrupy plastics on rotating rolls. Pairs of rolls create a layer of polymer of uniform thickness.

(4) is a head from which molten polymer is flowing. This head reciprocates left and right to cover full length of rolls. (6) is the nozzle which drops flowing polymer over the fine gap between rollers of pair (1). Melt layer coming out of role pair (1) is comparatively more viscous. Pair (2) make it still uniform and more viscous. Finally, layer of polymer is taken over by roller (3) which presses it over fabric sheet and smoothened by one or two more pairs of rolls further ahead. If needed, similar operation may be performed on the other side of fabric.

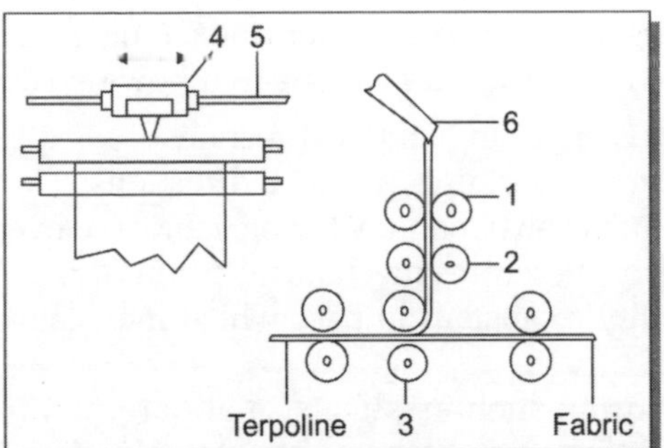

Fig. 2.100: Lamination system

FRP

FRP stands for fiber-reinforced polymer or fiber-reinforced plastics. Other terms as used in industry are:

RTP — Reinforced thermoset plastic
GRP — Glass-reinforced plastics
RTR — Reinforced thermoset resins

As the name suggests, FRP is polymer or plastics which are reinforced by long or short strands of fibers of materials such as glass and carbon. Jute and coconut hair are also used in hand lay up process. Diameter of fiber may vary from some micrometers to fraction of millimeters.

Epoxy, vinyl ester, polyester thermosetting plastics are generally used to produce FRP. These materials are in general termed matrix for the purpose of FRP manufacturing process. Glasses or carbon fibers are prepared in the form of woven sheets and forms. These materials are bonded with the matrix of polymer during molding process. Molding process is to compress the mix under specified pressure, temperature and time. After laps of time, mold or system is allowed to cool down and the product is taken out, inspected and auxiliary operations, if any are carried out. FRP is very useful material as it is quite strong, capable of replacing steel and other materials. It resists impact and dents. It is tough and flexible. With all its toughness and strength, it is a lighter material as compared to other materials such as steel, aluminum, etc. It very much resists corrosion even when exposed to sea water for long time.

FRP is used for a variety of products and parts. It is extensively used in automobiles, aircraft, marine, corrugated sheets, cubicles, shelters, boats, pipes, tubes, helmet, and a long list of other items. Manufacture of fiber fabrics is done in following two ways:

1. Two dimensions, in X and Y directions
2. Three dimensions, in X, Y and Z directions.

In three-dimensional manufacturing, fabric is woven through its thickness also, therefore, mechanical properties get improved, specially ability to sustain impact which may cause damage to FRP product.

Fiber performs such as sheets, mats are manufactured by weaving, knitting and stitching, etc. After performs are manufactured, molding process is done either by composite molding or molding. Composite moldings are carried out by:

- Bladder molding
- Compression molding
- Autoclave
- Mandrel wrapping
- Wet laying up
- Chopping gun
- Filament winding

Glass fibers are processed by:

- Fiber fabric
- Fiber mats
- Chopped fiber glass
- Short strands of glass fiber

Corrugated translucent sheets are classic example of fiber-reinforced polymer sheets. Photo 2.13 shows a corrugated sheet piece.

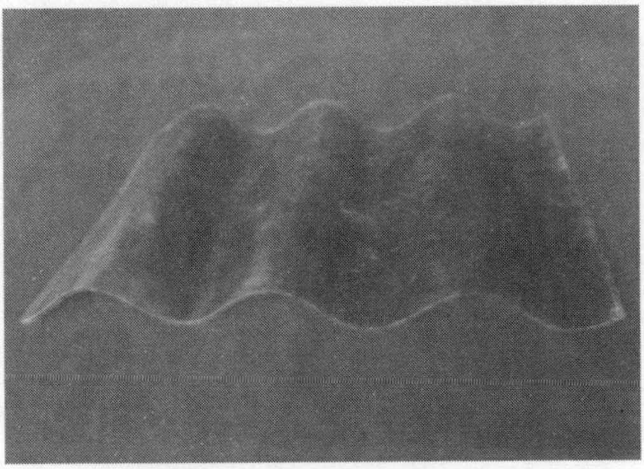

Photo 2.13: Corrugated sheet

Basics of hand manufacturing process is to lay fiber glass in the mold having corrugated shape as shown in Fig. 2.101.

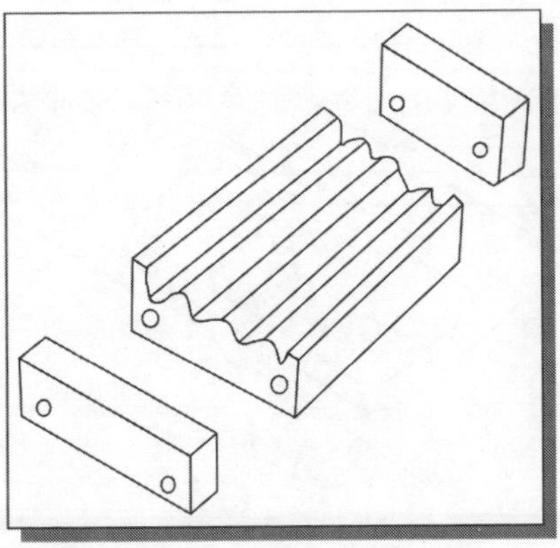

Fig. 2.101: Corrugated sheet die

Over the laid fibers, suitable matrix (polymer) is laid. Top portion of mold has matching corrugations and dimensions so that proper thickness is attained. Once the mold is closed, it is heated to appropriate temperature ranging from 180 to 220°C for one to two hours.

Composite Molding

Bladder molding is one of the techniques of composite molding. Figure 2.102 shows basic principle of set up.

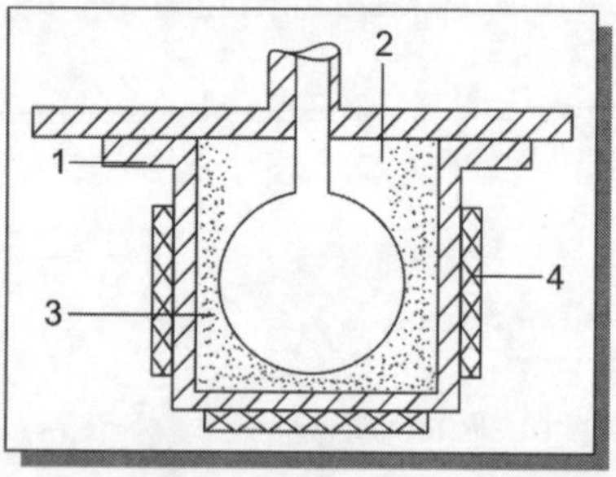

Fig. 2.102: Composite moulding

Figure 2.103 shows the cover of a machine which attains an ambient temperature of 80°C. Approximate dimensions of this typical RPF cover are also given. Principle of such molding is described with help of Fig. 2.102.

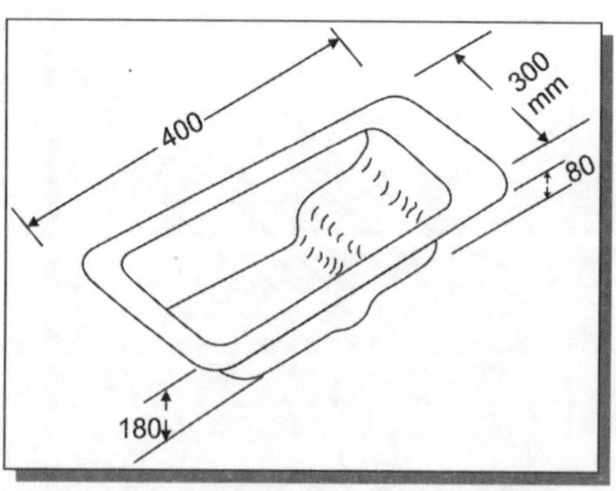

Fig. 2.103: Composite moulded component

(1) is metal mold cavity. Size and shape of cavity is so kept that molded component comes out with desired dimension. (2) is a mix of pre-form and matrix (polymer). (3) is a rubber bladder which is expandable by applying high pressure air. Consequently, mix is uniformly pressed against mold cavity walls. Temperature of mold is then raised to cause thermoset polymer matrix. Time required to achieve complete curing depends on type of mix, mold size and thermal properties of polymers such as epoxy, vinylester or polyester thermosetting plastics. Correct temperature and time are determined by initial practical trials. Parameters are recorded and repeatedly used.

Component shown in Fig. 2.103 may be produced by using mold of suitable shape and size.

PLASTICS WASTE MANAGEMENT

Materials such as metals, paper, wood, agricultural products are used as raw material to make variety of articles in the field of roads and building construction, ships, aircrafts, transport vehicles, pipes, tubes, machining parts, house hold items, kitchen wares, cloth, upholstery, science and technology related apparatus, medical equipment and a host of other items and durables. Almost in every production or manufacturing process, certain amount of material is not utilized. It may be due to inherent process design and it cannot be avoided. It may be minimized by improving upon process design and better performance of plant and machinery. Other type of waist may be due to negligence in handling, defective process, poor storing system and post process waste due to rejection of certain percentage of product on account of non-conformity with quality standards. This type of wastage is more where manual operations are involved. Percentage of wastage of material may range up to 10 to 15% or even more of total volume or weight of material used. Producers would not be able to afford such a high percentage of waste. Production may be stopped for taking corrective measures. Till this time, quite an amount of wastage has taken place.

In case of automatic plants, percentage of waste may be from zero to 1.5% or more. In high speed production plants, even 0.2% of wastage may mean piling up of waste over a short span of time. Hence producer would have to be very vigilant to take corrective action as soon as waste increases to cross zero percent allowable limit.

Management of plastics waste is necessary because most of plastics do not disintegrate over a long period of time. This pose a *threat* to environment. There are some plastics which are very slowly attacked by microbes. This phenomenon of microbe attack is so small and rare that it can hardly be termed as a natural means of plastics waste disintegration.

'In House', Industry

In injection molding plants, some wastages do take place in the form of granules, resins or powder spills and felling on ground. This generally happens in injection molding workshops where manual filling in machine hopper or filling of granules is done in hopper loader container. Such type of spillage may be kept to a minimum by careful manual handling. Such type of wastage is generally negligible in plants where supply of granules from silos to machine hopper is fully automated. If spillage is managed to be kept clean, free of dust and dirt particles, can be used without any deterioration in the quality of moldings. If spillage gets piled up over a period of time and gets dirty, use of such waste causes quality deterioration of molded articles. Other waste generation which takes place in injection molding process is due to purging of plastic melt, leakage between the nozzle and feed head of mold, rejected components with defective molding and runner, etc. When all such wastes get piled up to substantial quantity, then certain percentages are reused *in house* and remaining are sold to individual waste collectors.

In blow molding workshops, major portion of waste may be due to rejection of blow molded articles, specially during start up production. If rejected articles are blown from HDPE, PP or similar plastics, then they can be granulated in house and small percentage can be mixed with virgin plastics and fed to parison extruder of blow molding machine. In case of rejection of PET bottles, situation may be different. It is difficult to reuse rejects in house. Generally, pile up of such rejects is sold to scrap purchasers who are individuals or a group of them.

In extrusion shops, start up wastage is in the form of partly formed sheets, pipes, sections, strands extrusion, etc. These are generally cut, granulated and reused in house. Small quantities may become dirty and contaminated making it unsuitable for reuse. Such quantities are a waste and are generally disposed off. Similarly, there may be waste from other processes.

Waste management in industry is normally taken care of by industry itself as wastage means financial losses. In spite of all efforts by some small plastics process industries, wastage does not came down. It is because of the fact that machineries and molds inherently generate percentage of scrap ranging from 2 to 25% of shot weight. In case of injection molding, waste management effort on the part of such industry may be to get runner less mold made to replace old molds. It may need significant investment in mold making and good injection molding machines. But it may be worth doing as reoccurring saving may pay back investment. So in the long run, it may be a good effort for waste management.

Post Consumer Wastage

Migratory or mobile consumers are those who travel from place to place by train, buses, cars and other means of travel. During travel, the majority of them throw away trays, empty packets. They drink water, beverages and throw away bottles and polypacks. Consequently, waste plastics and paper get scattered here and there. At railway stations, a large quantity of waste gets piled up on the side of train where there is no platform, if it is not removed regularly by cleaning staff of railway station.

Situation of waste generation in localities of cities and towns are some what different. In few localities, there are fixed spots where garbage is thrown by residents. These fixed spots may exist by practice and habit of residents or fixed by municipality or corporations. Here the garbage may have variety of things like food and vegetable leftovers, broken pieces of plastics items such as children tricycle, basket, mugs, chairs, stools, cabinets, pipes, tubs, bottles and thin carry bags. On having a look on these spots in the morning, it may be seen that one or two small and young boys, men and even girls picking up pieces of their choice. Generally, these *rugs pickers* as they are called, leave thin carry bags, but lift other plastics, paper, rubber and metal pieces. On questioning, they say that thin carry bags do not yield them money worth their efforts in picking and carry them. A few rug pickers said that they hardly get rupees 4 to 5/kg of carry bags. Few years back, they used to pick up as at that time carry bags used to be of thick gauges. Needed less effort to collect a kilogram and got paid up to rupees 6 to 8/kg. Result of this situation is that carry bags of thin gauge and of a variety of shape and size are found scattered all over the places. These are the major causes of choking of small drains along lanes and roads. Ultimately drains are choked causing variety of problems and health hazards. Ultimately, rivers also become polluted.

Unorganized Collections

Very less educated or uneducated rug pickers sold their day's pickings to individual dealers who dictate their own terms of deal. Rug pickers hardly have any option other than to agree and get some money for survival. In turn, dealers supply collected plastics waste to parties who convert plastics waste to second grade granules. Only a few converting companies are located in some particular cities. Plastics wastes from various cities are transported by road transport, when full truck load is available for transportation. Its costs increases pressure on price negotiation as bulk market for second grade material is shrinking due to customer awareness and quality consciousness.

Industry-minded entrepreneurs and small industrialists do not invest in setting up even small output plant for want of availability of waste plastics, to fully utilize plant to full capacity.

Margins of profit are also very narrow, enough to discourage investment in the field of recycling of plastics waste.

Organized Collection System

Organized collection system starts with consumer attitude and civic responsibility. Consumer may keep waste plastics in their houses, business premises and office separately from other waste material.

Municipalities may develop plastics waste collection centers. Rug pickers may sell segregated plastics waste for a reasonably fixed rate in rupees per kilogram. Payment rate may be so fixed by municipalities that it motivates rug pickers to even collect carry bags.

Some NGOs may be subsidized and given *holidays* to set up plants for washing of plastics waste and then to sort out according to type of plastics. High impact polystyrene may be differentiated from HDPE. It may be difficult to differentiate between ABS and HIP because ABS waste must have lost its shine. Similarly, it may be difficult to sort out PVC waste.

Sorting mainly relies on sorters sense of tough and visual recognition. During washing process in tanks filled with water, heavier plastics like polystyrene, PVC, ABS, etc. would sink in the bottom of tank whereas lighter plastics such as LDPE, PE, etc. would float as their specific gravity is less than 1 (one).

There are fully automatic plants for processing waste plastics starting from waste to production of granules. Use of recycled plastics for durables is very limited. Roadside hawkers sell cheap items such as mugs, funnels, pipe connectors, spoons, soap cases and many other such items. Ultimately these items also become waste. So it continues cycle of virgin to waste.

Many thousand tons of plastics are produced in India yearly. About 20% of scrap is recycled. Maximum portion of remaining quantity must be utilized for generation of energy. Remaining quantities may be de-polymerized or used for land filling.

Following recycling technologies are placed as a plastics waste management effort:

- Feed stock regeneration by liquification
- Gasification or hydrogeneration

TEST

1. How many mostly used processes are there for converting plastics materials into products?
2. Briefly explain injection molding of thermoplastics.
3. Briefly explain how thermosetting plastics is processed to produce components?
4. What do you understand by extrusion process? Briefly explain with the help of sketch.
5. In injection molding machine, injection is carried out either by a plunger or a screw plunger. Explain why use of screw plunger is preferred?
6. What are the major classification of injection molding machines?
7. What is the difference between electromechanical and electrohydraulic injection molding machines?
8. Explain how value of mold locking force depends on projected area of cavities and runners of mold being loaded on machine?

9. What are *daylight* and mold opening stroke in an injection molding machine?
10. What is the difference between horizontal and vertical injection molding machines?
11. What are convertible or combination injection molding machines? Briefly explain advantages and disadvantages of such machines.
12. Name various types of injection molding machines and briefly explain each.
13. What are cavities, cores, ejectors in a mold? Briefly explain.
14. What are feed head, gate, runner and air vent in a mold?
15. Name various types of gates in an injection mold. Draw simple sketches.
16. List out molding defects in injection molding and causes.
17. What parameters are controlled in an injection molding machine?
18. What are the values of mold shrinkage, inch per inch for polystyrene, ABS and LDPE?
19. After injection, time is given for mold remaining closed. On what factors, this time depends?
20. How many types of extrusion processes are generally adopted in industry?
21. What is the difference between extrusion die design for pipe and sheet?
22. Briefly explain with the help of sketch difference between *in line* and *right angle* die head designs in an extruder.
23. What are various zones in length of an extruder screw?
24. What is LD ratio of an extruder screw?
25. What is the difference between nomenclature thermoforming and vacuum forming?
26. With the help of a simple sketch, explain process of vacuum forming.
27. What is calendaring process and its purpose?
28. What are calendaring rollers and their purpose?
29. How does uniformity in sheet thickness is maintained in extrusion of sheet?
30. What do you understand by finishing of plastics products, explain?

ASSIGNMENT

Note: Tick in box for correct statement and mark cross in remaining boxes

1. Injection molding machines are specified by
   (a) Dimensions .... □
   (b) Maximum shot volume .... □
   (c) Mould locking force .... □
   (d) All the above .... □
2. Which of the following statements is true?
   (a) Manually controlled injection molding machines give molding of consistent quality .... □
   (b) Semi-automatic injection molding machine cannot be manually operated .... □
   (c) Semi-automatic and fully automatic injection molding machines are similar in operation .... □
   (d) Vertical locking injection molding machines are suitable for insert molding .... □
3. Which of the following statements is true?
   (a) Low injection pressure is needed for plastics having high melt flow index (MFI) .... □
   (b) High injection pressure is needed for plastics having high melt flow index .... □
   (c) Fiber glass filled plastics need more injection pressure as compared to unfilled grade .... □
   (d) Components having thick ribs need more time in mold .... □
4. Which of the following values of mold temperature is generally maintained in actual practice for ABS?
   (a) 25°C .... □
   (b) 125°C .... □
   (c) 40°C .... □
   (d) 50°C to 70°C .... □
5. Which of the following is correct?
   (a) Submarine gate provides automatic degating .... □
   (b) Submarine gating gives clean degating .... □
   (c) Direct sprue gate is easily degated .... □
   (d) Hot runner system produces more scrap as compared to conventional runners .... □

6. In deep vacuum forming operation, which of the following is essential
   (a) Sheet may not be pressed between die face and pressure pad .... ☐
   (b) Provision of draw punch or core .... ☐
   (c) No provision of high pressure air as aid to deep forming .... ☐
   (d) Only air pressure and vacuum is sufficient for deep forming .... ☐
7. Bottles for drinking water packaging are generally produced from
   (a) LDPE .... ☐
   (b) HDPE .... ☐
   (c) Polypropylene .... ☐
   (d) PET .... ☐
8. Which of the following processes is suitable for producing bottles with precision thread formation?
   (a) One step blow molding .... ☐
   (b) Injection blow molding .... ☐
   (c) Rotational molding .... ☐
   (d) Molding from pre-form .... ☐
9. Bulk quantities may be produced by using which of the following methods in injection molding?
   (a) Many number of machines .... ☐
   (b) Multi-impression mold .... ☐
   (c) High speed fully automatic plant .... ☐
   (d) By operating plant on three-shift basis .... ☐
10. Calendaring is a process for
   (a) Producing circular article .... ☐
   (b) For cleaning surface of sheet .... ☐
   (c) For providing desired finish and texture to surface of sheet .... ☐
   (d) For making the thickness sheet or film uniform .... ☐

11 Which of the following components belongs to calendar?
   (a) Tie bars .... ☐
   (b) Chain and sprocket .... ☐
   (c) Rollers .... ☐
   (d) Rollers gap adjusting mechanism .... ☐

12. FRP is acronym of
    (a) Free rolling pinion .... ☐
    (b) Free radical polymer .... ☐
    (c) Fiber-reinforced plastics .... ☐
    (d) Fire-retardant plastics .... ☐
13. Which of the following processes is employed for producing fiber-corrugated sheets?
    (a) Extrusion .... ☐
    (b) Vacuum forming .... ☐
    (c) Glass fiber lay covered by epoxy and then pressed in mold .... ☐
    (d) Rotational molding .... ☐
14. Which one of the following is the method to fill machine hopper with plastics granules?
    (a) Granulation .... ☐
    (b) Manual loading .... ☐
    (c) Vacuum hopper loader .... ☐
    (d) Belt conveyor .... ☐
15. Plastics materials are supplied in which of the following forms
    (a) Clinkers .... ☐
    (b) Small oval balls .... ☐
    (c) Granules .... ☐
    (d) Powder .... ☐
    (e) Resins .... ☐

3 Plastics Testing Techniques

INTRODUCTION TO PLASTICS TESTING

From the onset of discovery, development and invention of plastics materials, predominant curiosity was to find out as to what a newly found plastics could be used for.

Gutta-percha is probably a classic example of first synthetic plastics which was tried for many uses. It was moulded in the form of sheet which was translucent. Scientist wrote to his friend about discovery of plastics *Gutta-percha* as a translucent material. Scientists must have been asked *how much translucent, can I see through it, can I read*, etc. Gutta-percha sheet must have been subjected to bending to get a *feel* as to how strong it is. It must have been subjected to burning trials (a crude test from today's standards). They must also have tried to see the effect of sunlight, acids and alkalies, etc.

All above trials were mostly descriptive. No quantitative reports were made because by early 19th century scientific test methods and equipment were not sufficiently developed.

Significantly, with continuous development in science and technology, various types of plastics such as polyethylene, cellulosics, etc. came into existence. Use of plastics found its way for a variety of purposes.

Around year 1942 and World War II, need of qualitative and quantitative description was felt a great necessity. Almost all developed and developing countries took initiative to develop various tests for polymer products produced from various plastics. This phenomenon occurred due to various factors, predominantly pressure from military to develop plastics so that economical and better performing items such as tarpaulin, ropes, jar-cans, buoys, sheets, pipes, etc. could be produced to suit battlefield environment. Growth of industry further demanded specific quality norms. All these pressures led governments of various countries to establish organizations and departments to take care of development of quality norms for various thermosetting and thermoplastics. Some countries concentrated in developing and enforcing quality norms in their own land. But with fast developing globalization, plastics materials of various types and grades started moving from one country to another.

For example, during mid nineteenth century, number of plastics such as high density polyethylene, nylon, polypropylene were imported from other countries to India. All these activities led the development of national and international agencies, governmental organizations and NGOs to develop standard test methods to accurately convey quality description in the form of standards among producer, seller, buyer and end user.

A few examples of present organizations are mentioned below.

ASTM : American Standard Test Method
DIN : Deutsches Institut Für Normung
BIS : Bureau of Indian Standards
BS : British Standards
JIS : Japanese Standards Association
ISO : International Organization for Standardization
EN : European Standards

ASTM

American test methods as we know today were developed out of a necessity around end of 18th century or beginning of 19th century. At that time, some steel rails were breaking. It was discovered that reported failures were due to variation in manufacturing technique, composition and grain structure. This led to critical studies and framing of standards.

ASTM is one of the most popular test methods which cover a very large number of materials, finished products, constructions, public services and processes of manufacturing. A very comprehensive website is available on internet.

http://www. ASTM. org

There are steps and facilities to search information of your choice. For example, if one needs to know availability of ASTM documents for plastics, it can easily be searched. On selecting melt flow index, for example, test method document number ASTM D1238 would be displayed. Detailed description of test method and equipment is not generally provided on website. *Advance search* may provide some details. One can also search from ASTM documents in print or electronics may be purchased. This facility is also available on website.

http://www. wikipedia. org.

DIN

DIN is an acronym of *Deutsches Institut Für Normung*. In English, it is German institute for standardization.

It is also a Logo of the German institute for standardization which is located at Berlin. It was founded in 1917 with its name as NADI, *Standardization Committee of German Industry*. In 1975, it was renamed as Deutsches Institut Für Normung and DIN.

DIN EN is used for German edition of ISO (International Organization for Standardization). DIN standard covers almost every aspect of technology. There are standards in thousands. Selection of particular standard is made easy on visiting related website on internet. Any search engine such as Google may be used to get to website for DIN. Wikipedia the free encyclopedia on internet may also be visited. There may be a long tabulated list of DIN standards with the topics. Going through the list is quite time consuming. So there may be an option for search. One may write on search bar the desired topic and by clicking *search*, desired standard may be seen on monitor screen.

For example, *polymer blend thermoplastics* would present

DIN 16780-2

01-Oct-1990

Till writing this book, detailed description of standards could not be found. It might be necessary to purchase particular standard.

BIS

Indian standard institution (ISI) was founded is 1947 with the objective to provide a platform for Indian industries, producers and consumers to follow uniform norms of quality and standard. Over a period of time, a reduction in losses to industry and trade was obvious. Consumers also become more satisfied with goods carrying ISI marking. It was and is due to the fact that standards were set by committees comprising of highly experienced experts drawn from institute, industries, government organization, NGO and members of public. Standards set by ISI were voluntarily followed by producers and service providers. For example, electric ON/OFF switch become loose thus give flickering contact, if they are not manufactured as per ISI standards. Switches having ISI marking give a satisfactory service, say up to 10,000 (ten thousand) switchings.

Name of Indian standards institution was changed to Bureau of Indian Standard, who's acronym is BIS. Its activities are not confined to only developing standards. Other activities are:

- Educating people about various activities of BIS through lectures, seminars, demonstrations, propagation of information through print and electronic media
- Guidance to industries in taking improvement measures
- Testing of materials and products
- Many other activities

BIS is also member of ISO.

ISO

ISO is acronym of International Organization for Standardization. Standard organizations of most of the countries are the members of ISO. It is almost a must because of increasing international trade and exchange of technology.

ASTM, DIN, BS, JIS, EN, etc. are members of ISO.

MECHANICAL PROPERTIES

Examples of few items would be taken up to discuss about their use in actual practice and related mechanical properties.

Ropes made up of plastics are of various diameters and strands. These are subject to twist, bend and pull. If two ropes of similar dimensions are made with different plastics, then the performance of these ropes would be different. A rope made of LLDP would get elongated under tension while the other made of polypropylene would get nominally elongated. It is because of the fact that tensile strengths of two plastics in question are different which are as follows:

LD polyethylene .... 3000–6500 psi
Polypropylene .... 4500–6000 psi

Designer of a rope would need to know the value of tensile strength of plastics to calculate the diameter and number of strands for a rope to sustain ultimate tensile loads safely.

Other property parameters such as elongation at break percentage, may also be needed to be known.

Straps

Straps of a wrist watch are made from leather. But nowadays plastics straps are commonly used. These straps in use are hardly subjected to significant value of tension.

These straps are subjected to quite a number of bendings. Some straps show sign of cracking quite early, whereas others do not show sign of cracking for a long time. This is basically due to the fact that flexural modules of plastics might be different.

This property is expressed in 10^3 psi.

Since values of flexural modules are generally given by plastics producers in their technical brochure, designer may choose an appropriate type and grade of plastics.

Packing Straps

Since last many years, plastics straps have become widely used alternative to steel straps. Steel straps are still used for very heavy duty work, such as bundling of iron rods/bars as used in building construction. Plastics straps are abundantly used for binding big bails of cloth, jute sacks, terpolene, etc.

Binding straps, in the first place, are subjected to high tensile force and during transport and unloading, straps are further subjected to sudden tensile forces. Therefore, these straps are extruded by such polymers or copolymers or fiber-filled plastics that their tensile strength is of high order. Accordingly, thickness and width of straps are calculated. Standardization provides same property standards to end user, designer and producers. Hence there is hardly any chance of disagreement over property of plastics by which a particular strap is made of.

Plastics Sheet

Extruded films and sheets of a variety of thickness, width, plastics, laminated plain, laminated woven are available in market. Generally, these sheets are used for packing, covering and laying in ponds and canals. It is observed that some films are quite strong against tearing, rupturing or expanding. All these depend on mechanical properties of plastics materials from which films are extruded and calendared. Performance mostly depends on the following properties of polymers from which sheets and films are produced.

- Tensile strength
- Percentage of elongation at break
- Flexural strength

Sheets and films producers select polymers of particular properties including mechanical, keeping in view the quality demands from customers (end users).

Travel Suitcase

Travel cases are subjected to harsh handling, such as drops from height, scratches and loads on them. Sometimes travel cases are over-packed, causing tops to bulge out.

In market, a variety of cases are available, which vary in dimensions, shapes, fittings and type and grade of plastics.

Branded travel suitcases are generally molded from plastics which have following properties:

- Tough
- Hard
- Good flexural hardness

Designers of travel suitcases study mechanical properties of various types and grades of plastics and workout manufacturing cost. Cost and quality factors are then reviewed by authorities concerned.

Finally, polypropylene or copolymers of polypropylene may be chosen as compared to HDPE. A better plastics may also be selected which is fiber glass-filled polypropylene and is much tougher, strong and scratch-resistant.

Chairs

Plastic chairs are generally subjected to rough handling during transport and use. Choice of plastics is little difficult as the predominant consideration is material cost. Plastics chairs are designed with a lot of ribs and legs with sections to achieve sufficient strength even with low toughness and compressive strength plastics. High density polyethylene may be chosen.

Documentation of properties by plastics manufacturers enable the designer to compare values of various mechanical properties and to choose one of his choice.

Small and Big Containers

Many small containers in round and rectangular shapes are molded out of clear general purpose polystyrene. These are low-priced items having clear view of small items such as buttons, U clips, erasers, screws, electronic components, etc. kept inside them.

In terms of izod impact, its strength is quite weak. It may break by falling, impact or exertion of load. In spite of this mechanical drawback, these are used at office table, on storing racks and almirahs.

Vehicles Seat Covers and Linings

Materials used for seat covers and linings are generally extruded and calendared sheets. These are flexible and strong sheets generally produced from PVC. Laminated sheets of copolymer are also used.

Tensile strength, flexural modulus are important properties of polymers which are taken into consideration while selecting a particular type and grade of plastics.

THERMAL PROPERTIES

Some of thermal properties of plastics are as follows:

- Vicat softening temperature
- Deflection temperature under flexural load
- Thermal conductivity
- Specific heat
- Relative temperature index, mechanical
- Coefficient of linear thermal expansion

These thermal properties attain significance, if product has to perform under raised environmental temperature such as under the bonnet lining, outer panels of space crafts, technical components used for oven construction and other equipments. Knowledge of thermal standards of polymer is also important when designing inside insulation lining of automobiles, refrigerators, cool chambers, water coolers, etc.

ASTM tests for above properties are as below:

- Vicat softening temperature
  Rate A D1525
- Thermal conductivity–C177
  - Specific heat...................... D2766
- Relative temperature
  Index .... {UL7468}
- Coefficient of linear
  Expansion........................ D696

Values of these tests may vary from type and grade of plastics produced by various producers. Since the basis of determining the properties is same standard test method. Designer may choose a particular plastics with suitable properties as per declared values measured by standard test methods.

PERFORMANCE PROPERTIES

Performance of polymers may be considered for various functions, namely:

- During conversion process
- During use of product
- Utility time span
- After performance, life of product

Conversion process means that plastics are converted to desired product. Products may itself be a final product like injection molded items, bottles caps, etc.

Lay flat sheets, films, calendared sheets may not be final products because some items may be produced out of these sheets like raincoat, bags, lining and covers. Pipes and tubes are produced by applying extrusion process. Pipes and tubes are generally final product which may be used for many purposes. Performance of plastics during conversion process need certain properties to perform well. There is no doubt that performance of plastics is related to process system, design and parameters.

At present, it is presumed that all process parameters are ideal or favorable. It is only the plastics/polymer performance which is taken into account here. In case of injection molding, properties of plastics should be such that it provide following performances.

- Ease in cavities filling. This means that high values of injection pressure is not needed. Backup pressure (packing pressure) is needed for a short period of time.
- Cooling of molded items in cavities should be fast. This means that plastics should give out heat to cavity which has cooling system while cooling is taking place, no sink marks get created. Flow properties should be such that no flow marks appear on component surface.

Similarly in case of blow molding, properties of plastics should be such that it provides following performances:

- Parison should not get elongated by its own weight
- Parison walls should be free from air or gas bubbles.
- Does not have sticking tendency to cold or warm parison cutter.
- Homogeneity in the plastics of parison so that wall thickness of blown item is uniform for a particular section of product.

Assuming that all process parameters and die head setting are favorable, then the quality of performance may be attributed to characteristics of polymers.

In case of thermoforming, properties of plastics should be such that it provides following performances:

- Mass of sheet gets heated to achieve pliability in short period of time. No excessive heat and time should be necessary to make the sheet pliable to suit vacuum forming.
- On thermoformed portion of part, there should not be sign of strains, generally visible in form of minute cracks. Strength of vacuum formed parts should remain intact by virtue of properties of plastics, tensile strength, flexural strength, etc.
- Fast dissipation of heat to thermoforming mould.

Calendaring properties of plastics should be such that it provides following performances:

- Should not have sticking tendency to rollers sheets while passing through gap between rollers get roll pressed to attain desired thickness. Consequently sheet elongates and widens. As a result of this action, chain structure of polymer should not get disturb to undesirable level.
- Quality of polymer should be such that surface of sheet attains surface smoothness and shine of highly polished and mirror bright surface.

- In case of texture, design on rollers, films should attain minute details of texture and should retain while cooling of sheets to ambient temperature takes place.

During Use of Product

Performance properties of plastics should be such that a product during use exhibits following properties.

- Retains its appearance for the time period for which product is expected to remain usable.
- Many a products have esthetic printed information or logo on plastics surface. So the surface of plastics product should be such that adhesion of inks or paints should be durable.
- Migration of additives and dyes in plastics should be under permissible limits.
- Wherever necessary, plastics should be so chosen that it is fire retardant.
- Chances of season cracking should not be there for product normally used out door and exposed to weather conditions.

Utility Time Performances

New HDPE bucket has surface luster and flexibility which can be felt by flexing it. When this bucket is used for a long tisme, it starts getting brittle. With passing time, brittleness increases to such an extent that portion of wall breaks, if subjected to impact or brim of wall gets bent. This example provides concept of utility time performances. Time-dependent performance depends on factors like reduction of modifiers, effect of sunlight, use of various liquids, mechanical strain on polymer chain structure, etc. These effects are mostly very slow. Deterioration in performance quality may take years and even decades. Sometimes it is found that signs of degradation start appearing within year's time. Why such a wide variation? It all depends on the types and grades of polymers. In case of bucket example, it can be argued that generation of brittleness may be due to deficiencies in mould design and processing conditions. These deficiencies mostly resulted in development of cracks over a period of time and not noticeable brittleness. So, it may be said that useful performance life of plastics product is also time dependent.

PROCESSING AND FLOW PROPERTIES

Following properties of polymers predominantly affect processing and flow properties:

- Melt flow index
- Specific gravity
- Processing temperature
- Percentage of fillers like glass fibers, carbon
- Optical properties

In various standards, melt flow index is specified as grams of polymer flow through a specified diameter hole, under specified temperature and pressure in 10 minutes. A type and grade of plastics may have an MFI value as 23 grams per 10 minutes. A different grade of same type of plastics may have 10 grams per 10 minutes. It is, therefore, expected that flow of plastics having MFI as 23 would be easier to fill up cavities in injection moulding process.

Specific gravity is determined by test methods such as ASTM D 792.

Any polymer which is unfilled would have a low specific gravity as compared to glass or ceramic filled grade.

Processing of unfilled polymer requires less injection pressure, melt temperature and filling speed as compared to filled grade.

Processing Temperature

Processing temperature varies from type to type and grade to grade of plastics. Polymers with low processing temperature may have following advantages over high processing temperature grades.

- Less moulding cycle time
- Less heating power consumption
- Less sink marks
- Less chances of plastics degradation while polymer resides in plasticizing cylinder.

Percentage of Fillers

It is already mentioned that processing parameters vary with percentage of fillers. Some types of plastics have grades of various types of fillers and of varying percentage. Fillers are introduced in virgin polymers to modify properties of polymer, specially mechanical properties.

Table 3.1 provides typical idea of fillers and percentage in few polymers.

Table 3.1: Typical idea of fillers and their percentage in few polymers

Polymers	*Fillers*	%
1. Polystyrene	Long and short glass reinforced	20%
	Carbon fiber	20%
2. ABS	Aluminum flakes	40%
	Carbon fiber	20%
	Glass fiber reinforced	20 to 30%
	Long glass fiber reinforced	20 to 40%
3. Polyamide	Glass fiber reinforced	35 to 45%
	Mineral and glass fiber reinforced	40%
4. Polycarbonate	Glass	10–30%
	Long glass fiber reinforced	20 to 30%

OPTICAL PROPERTIES

Optical properties of glass or plastics, etc. include the following

- Refractive index
- Light transmittance
- IIaze
- Transmittance of ultraviolet rays

Refractive Index

When a ray of light travels from one medium to another, it deflects towards or away from perpendicular to the meeting surfaces of two media. It is illustrated in Fig. 3.1.

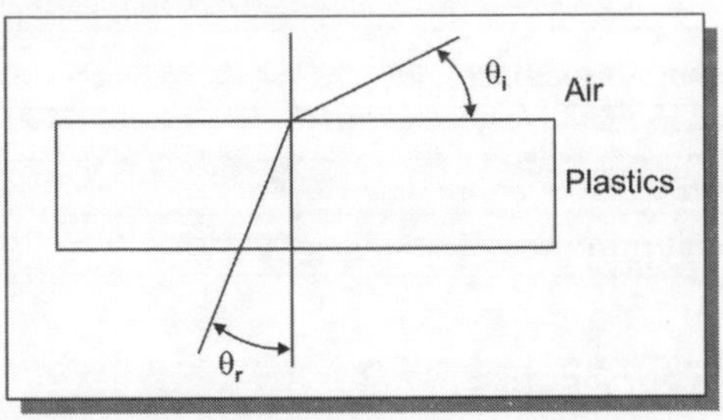

Fig. 3.1: Refractioning plastics

θ_i is the angle between incident ray and perpendicular. It is called angle of incidence. θ_r is the angle between deflected ray and perpendicular. It is called angle of refraction. Refractive index is the ratio of angle of deflection to angle of incidence.

$$\partial_i = \theta_r/\theta_i \qquad \text{which is a number}$$

Light Transmittance

Suppose a beam or ray of light is aimed at a sensitive optical device which accurately senses and displays the amount of light falling on the device sensor. It may be displayed in candle power or lumens. Let there be an experiment set up where the intensity of falling light is calibrated to display it as 100%. Now strip or block of clear glass is brought in the path of light bean. It would be observed that digital display is shown as 96%, for example. This means that there is drop of 4% of light. It would be then said that light transmition of that particular sample glass is 96%. Similarly, various types of clear plastics have different values of percentage of light transmittance.

Percentage of light transmittance in plastics also depends on how well smooth and polished surfaces of sample plastics are. BIS, ASTM, etc. specifies test methods which include specified quality of surface finish of sample.

Percentage of light transmittance for clear plastics is generally given in technical brochures of various plastics manufacturers. To give an idea of variation in percentage of light transmittance, typical values are given in Table 3.2.

Table 3.2: Light transmittance

S.no.	*Polymer*	*Percentage of light transmittance*
1.	Polymethyl methacrylate (PMMA)	97–96%
2.	Polycarbonate (PC)	95%
3.	Polyethylene terephthalate (PET)	95–96%
4.	General purpose polystyrene (GPS)	94–92%

ELECTRICAL PROPERTIES

Plastics is abundantly used in manufacture of electrical gadgets and equipment. It is necessary that performance of parts and gadgets in electrical environment should be safe and provide desired performance. There are number of tests which are meant for determining the suitability of a particular type and grade of plastics. Some of these test are:

- Dielectric constant
- Dielectric strength
- Dielectric dispersion
- Surface resistivity
- Volume resistivity.

Standard tests are developed on international level by testing agencies like ISO, ASTM, BIS, etc.

Table 3.3 provides typical value for some plastics as to give an idea of various test methods.

Table 3.3: Test values for some plastics

Tests	*Test method ASTM*	*ABS*	*Cellulose acetate*	*HDPE*	*Polystyrene*
Dielectric constant	D 150	3.2–3.3	~ 5	2.3–2.4	2.4–3.1
@ 1 MHz	D 2520	–	–	–	–
Dielectric strength	D 149	20–25	11	22	20
KV mm	–	–	–	–	–
Dissipation factor	D 150	–	0.06	–	0.0002
1 KHz	D 2520	–	–	–	–
Dissipation factor	D 150	0.02	–	$1–10\times10^{-4}$	–
1 MHz	D2520	–	–	–	–
Surface resistivity	D 257	–	–	10^{13}	–
Ohm/sq					
Volume resistivity	D257	$>10^{15}$	5×10^{12}	$10^{15}–10^{18}$	$10^{15}–10^{18}$
Ohm/cm					

Source: http"// www.professionalplastics.com

Dielectric Constant

Capacitors of various values, shapes and sizes are used in electric and electronic circuits. Performance of a capacitor depends on many factors. One of them is dielectric used in a capacitor.

Dielectric is an insulating material between two electrodes. This material gets polarized in an electric field. Consequently, an electrode is charged. This charge does not flow through dielectric material. High polarization is expressed by a number called dielectric constant.

Plastics is also a dielectric. Table 3.3 gives typical values for few types of plastics. Higher the value better is the performance.

Dielectric Strength

Dielectric strength of dielectric, say plastics is the maximum field strength which it can stand without breaking down. Hence, the performance of a capacitor would be better with a dielectric of high strength.

Dielectric Dispersion

Dependence of permittivity of a dielectric on the frequency of applied electric field is called dielectric dispersion. Performance of equipment working on high frequency also depends on dielectric dispersion.

Surface Resistivity

Electrical current experiences a resistance in the flow on the surface of plastics from one place to another. Figure 3.2 shows a typical set up for measuring surface resistance.

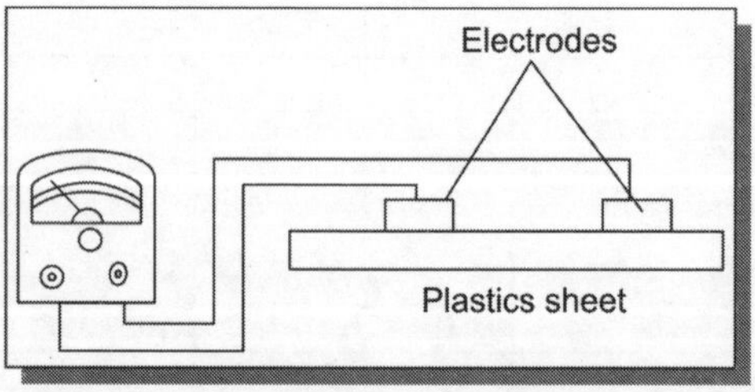

Fig. 3.2: Surface electrical resistance

Surface resistance is tested according to test method ASTM D257. Typical values of surface resistance range from 10^{14} to 10^{18} (ohms per square cm) ohms for intrinsic polymers.

Safe performance of an outer plastics components in electrical environment depends to a great extent on surface resistivity. It is worth noting that atmospheric humidity affects surface resistivity, hence performance of plastics components or part.

Volume Resistivity

It is the resistivity in ohms per centimeter of thickness of plastics material. It is the resistance to the flow of electrical current through the mass of plastics and NOT surface. Recommended test for measurement of volume resistance is ASTM D257.

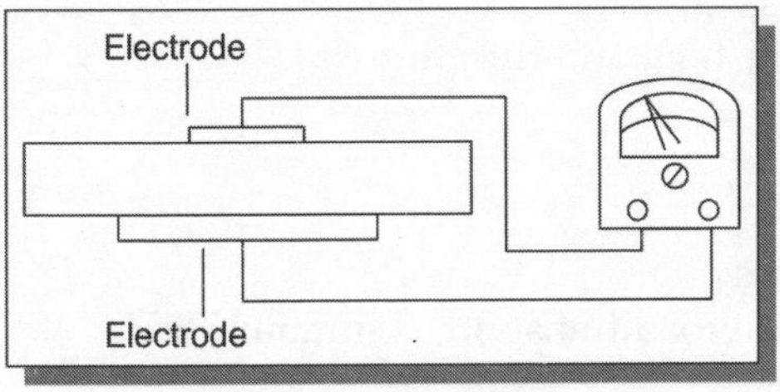

Fig. 3.3: Volumetric electrical resistance

PRODUCT TESTING

Modern consumers of raw material, parts and goods are quite conscious of quality for which they spend money. Quality in terms of specific aspects is advertised and documented in product leaflets by manufacturers of goods. Manufacturers are aware of the fact that if their products are not up to the mark in terms of quality aspect, service and safety, then the market share of their brands of product would erode and ultimately may be wiped out from market place. This situation may not arise, if products are thoroughly tested by nationally or internationally established test methods. Few nationally and internationally test method are mentioned below.

- European Standard
- GB Standard (Standardization Admistration of China—SAC)
- ISO (International Organization for Standardization)
- ASTM (American Standard Test Method)
- BIS (Bureau of Indian Standard).

There are a wide variety of raw materials, products and services for which number of test method run into thousand. Varieties of products which may be subjected to tests are categorized in following manner.

- Accessories
- Cosmetics
- Food
- Footwear
- Health care and makeup products
- Home appliances
- Electronics
- Home furnishing and furniture
- Leather
- Luggage
- Textile
- Toys, games
- Services like road transport, communication
- Health care
- Games products, etc.
- Raw material, plastics, paper, wood, etc.

Manufacturers and producers of any of the above products may conduct standardized test provided as they have well-equipped and managed laboratories. Having in house testing facilities may be quite expensive and only corporate houses may afford to have testing facilities.

There is another aspect to it. Test certificate from producer or manufacturer may or may not satisfy consumer. It is, therefore, a general practice to get the product tested by nationally or internationally established organizations like Bureau of Indian Standard. There is a provision that government recognized organizations provide test markings such as ISI on the product to indicate that product has passed quality and safety tests. Generally, consumers have certain confidence that ISI marked products would perform as stipulated.

In almost all advanced and developed countries, there are test houses which have facilities to carry out a number of specific tests. Many of these test houses enjoy confidence of consumers. A few are mentioned below.

India .... Shriram Institute for Industrial Research
USA .... Intertek Plastics Technology Laboratories
UK .... Ashwell Hertford, ref: test house directory
Germany .... MPA RNW
Japan .... Yasuda Seiki Seisakusho Ltd.

Products in market place may be exclusively of plastics or a combination of components or parts made of various materials like metals and non-metals. Irrespective of materials of construction, product must perform according to stipulated norms namely extent of effect of jerks and vibration, impact, exposure to sunlight, exposure to humidly and sea water, weightlessness in outer space, proximity to acids, bases, gases, petroleum products, organic matter, entrance temperature, ambient temperatures, libration of gases and fumes on burning, microwave energy, ultrasound waves, electrostatic charge and many others.

All tests are designed and implemented after thorough deliberations by subject matter committees which consist of members from technical institutions, producers, test houses, consumer forum and other relevant organizations. During the process of designing tests, details of testing equipment are also finalized.

In many cases, manufacturers or producers improvise their own testing systems. This arrangement is used in day-to-day checking and testing of product being manufactured. Many a times, unacceptable quality performance is detected. Had it not being detected, a large number of components of defective quality would have been produced. Further, if sample were sent to test houses for certification then might have been rejected or failed. Consequently; wastage of time, effort and money would have taken place.

A specific example of *in house* testing is provided in following paragraphs to give an idea as to how in house quality testing of product is carried out.

According to standard specification, a switch of a flashlight should function correctly for twenty thousand strokes. Correct functioning means that ON/OFF of flashlight should take place without flickering of light or loose contact. A motorized jig was designed and made to give any number of strokes to switch. Figure 3.4 shows a typical arrangement of jig.

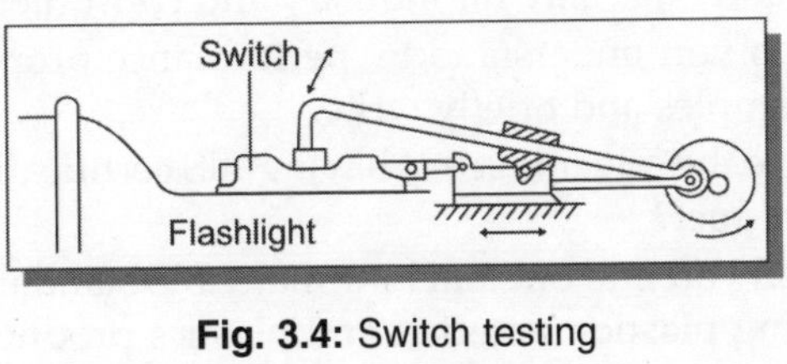

Fig. 3.4: Switch testing

After every five thousand strokes, functioning of switch was examined and observation recorded. Another test performed in house is life testing of miniature lamp as used in a flashlight. Its life expectancy is normally specified as 40 hours, when operated at a specified voltage. During production, it is highly undesirable to wait for 40 hours for getting life test results as during that period thousands of lamps would have been produced. In case life of lamps comes out to be less, then the lot produced may be labeled as defective. This problem is overcome by devising accelerated life test. About ten lots of lamps are picked up from production line. Each lot is divided into two. Set of one lot is tested for life on specified voltage, whereas remaining lots are tested on higher voltage so as to get results within two hours. Results of both lots are tabulated and graphs prepared. From *in house* testing point of view, accelerated test is good enough to feel safe in continuing production.

TEST

1. Around what period of time, plastics was developed and exhaustive work of further development and use was done?
2. Give classic example of synthetic plastics which were initially developed around mid nineteenth century (around 1940).
3. How does second world war had caused speedy development of plastics and associated technology?
4. Write down full form of these acronyms: BIS, ASTM, BS and JIS.
5. Write down website addresses of BIS, ASTM and BS for retrieving information.
6. ISI is the initial acronym of Indian standards institution. Around what year was it changed to BIS, Bureau of Indian Standards?
7. What are various activities of BIS for the benefit of general public and specially for industry and consumer?
8. What do you understand by performance properties? Give few examples and briefly write.
9. What are the advantages of having BIS certification marking on a product?
10. How can you find out that how many BIS standards are there regarding plastics materials and plastics products?

ASSIGNMENT

Note: Tick in box for correct statement and mark cross in remaining.

1. Which of the following is performance properties?
   (a) Mouldability ☐
   (b) Tensile strength ☐
   (c) *Life* of a product ☐
2. Which of the following is an international organization?
   (a) JIS ☐
   (b) BS ☐
   (c) ISO ☐
3. Which is the predominant property needed for packing straps (for binding crates, etc.)?
   (a) Tensile strength ☐
   (b) Melt flow index ☐
   (c) Izod impact value ☐
4. Which is the predominant property needed for items like wrist watch strap?
   (a) Hardness ☐
   (b) Vicat softening point ☐
   (c) Flexural tensile strength ☐
5. Following plastics have different cost per unit weight, which one is generally most expensive?
   (a) Phenol formaldehyde ☐
   (b) Polycarbonate ☐
   (c) Polyamide, glass filled ☐
6. Following plastics have different cost per unit weight, which one is generally less expensive?
   (a) Phenol formaldehyde ☐
   (b) PVC injection grade ☐
   (c) LDPE ☐
7. Buckets, chairs are generally moulded in which of the following plastics.
   (a) HDPE ☐
   (b) LDPE ☐
   (c) Polypropylene ☐

8. Phenolic plastics are suitable for which of the following purposes?
   (a) Building, electrical, electronic .... □
   (b) Vehicle bumpers .... □
   (c) Chemical storage tank .... □
9. Good quality travel suitcases need which of the following qualities?
   (a) Toughness .... □
   (b) High shore hardness .... □
   (c) High density plastics .... □
10. Which of the following qualities you consider important for plastics sheet and film?
   (a) Good light transmittance .... □
   (b) Much hardness .... □
   (c) Good tear strength .... □

4 Machine Maintenance

MAINTENANCE METHODS

Meaning of word maintenance as given in dictionary is upkeep, continuance protection. Taking clue from these meanings, it may be said that machine, maintenance implies that machine should perform as it should be. There should also be reliability in the performance of machine and that too for a longer period of time. If a machine manufacturer is asked as to what is the expected life of thc machine, then the most lightly answer would be that it all depends on how machine is operated, cared and maintained.

Machines are built up by assembling a variety of machine elements like shafts, gears, cams, livers, bearings, sliding guide ways, spline shafts, tie rods, hydraulic and pneumatic gadgets, fasteners, etc. Most of these machine elements are assembled together with pre-determined tolerances. Assembly of a journal and bush bearing is a classic example for analyzing its performance. A typical analysis of journal and bush bearing assembly is given below.

Figure 4.1 shows schematic assembly of journal and bush bearing for which tolerance is to be decided. Only designer of mechanism can decide as to what tolerance is to be given. Decision depends on the type of fit which is to be provided to assembly of shaft and bush.

Table 4.1 provides typical values of various fits.

Suppose designer decide to provide sliding fit then hole would have to be provided H7 tolerance and shaft would have to be provided g6 tolerance. Value of H7 or g6 tolerance depends on nominal diameters of hole and shaft. There are tables available from which recommended value of H8 or f8 for various nominal

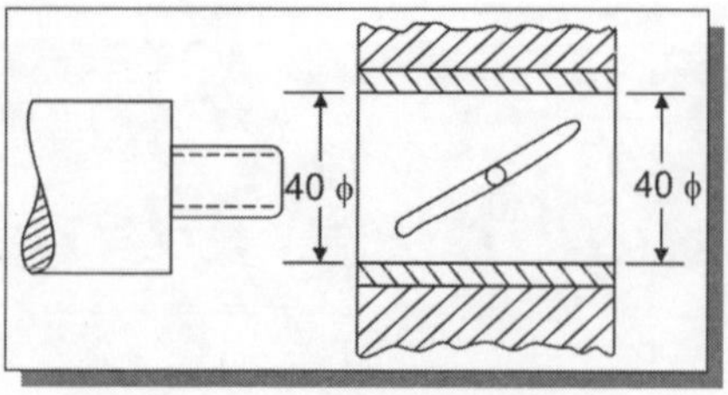

Fig. 4.1: Hole and shaft fits

Table 4.1: Fits and tolerances

S. no.	*Description*	*Hole tolerance*	*Shaft tolerance*
1.	Loose fit	H 11	c11
2.	Free running	H9	d9
3.	Easy running	H8	f8
4.	Sliding fit	H7	g6
5.	Close clearance	H8	f7
6.	Location clearance	H7	h6
7.	Slight interference	H7	k6
8.	Transition fit	H7	n6
9.	Press fit	H7	p6
10.	Medium fit	H7	s6
11.	Force fit	H7	u6

diameters can be found out. In case of Fig. 4.1, nominal diameters of hole and shaft are 40 millimeters.

- Value of H7 for 40 mm nominal hole,
  Typically is .... +0.025, +0.0
- Value of g6 for 40 mm nominal shaft,
  Typically is .... –0.009, –0.025

If the hole size is on the upper side of tolerance, +0.025 and shaft size is on the lower side of the tolerance, –0.025 then the total clearance between hole and shaft would be (0.025 + 0.025) 0.05 mm. It is the designed limit of clearance. Any increase in clearance may need some remedial action, namely replacement of bush bearing. It is a maintenance action.

Plant and machine maintenance may be categorized in following manner:

- Routine maintenance
- Scheduled maintenance
- Preventive maintenance
- Breakdown maintenance.

Routine Maintenance

Routine maintenance is normally carried out before start of shift and after the production shift is over. This maintenance is often done about one hour before the start of shift. First of all machine is cleaned. Cleaning of machine means removal of particles or broken chips of materials which are fed to machine, cleaning of leaked out fluid, etc.

Checking of various meters and control values for their proper functioning. Checking of three-phase or single-phase voltage supply, high and low pressure air supply, if machine is connected to vacuum line, then to check the quality of vacuum.

In routine maintenance, most important is to check if all safety provisions are functioning well. There are some assemblies where due to vibration, etc. some of nuts, bolts or screws become loose. So in routine maintenance such parts are checked for any looseness. If found loose, are to be tightened.

Scheduled Maintenance

Plants and machineries which are in regular operations are subjected to wear, tear and formation of rust or scales. In many machines, certain parts get intermittently heated while holding components and passing through burner flames or heating ovens.

Figure 4.2 shows turret (1) of a machine which is fitted with 18 sets of holding fingers (2) and (3).

These fingers hold pairs of wires (5). On the top surface of fingers rests a glass bead (7). Turret keeps on intermittently rotating. Set of holding fingers stops for few seconds in front of a burner flame(4). Flame keeps on impinging on surfaces of fingers. Intermittent heating of fingers throughout 8 hours shift adversely affect the surfaces of fingers, specially top surfaces where glass bead rests. Over a period of time, surfaces of fingers get pitted and rusted, consequently a stage comes, when holding of wires becomes defective, also at unloading station pair of wires with fused glass bead sticks, causing the *mounted bead* to de-shaped.

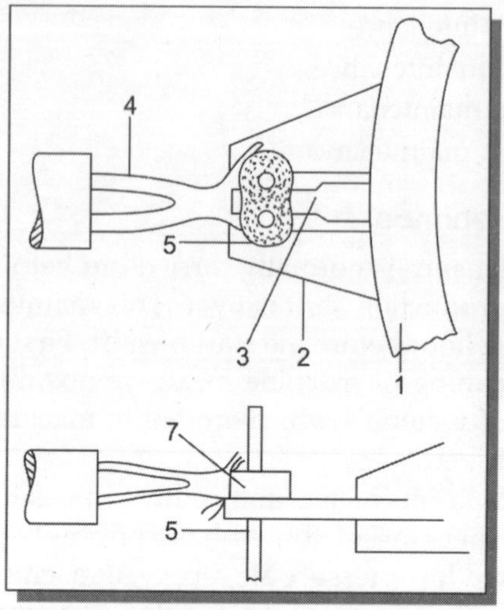

Fig. 4.2: Heating defect and glass bead

Over a period of time, maintenance technicians and workman know that defects start developing after about 20 days of working. So maintenance supervisor decides that every 15 days of working, sets of fingers would be replaced by good sets. It is normally done on a non-production day. This is called *scheduled maintenance*.

Another classic example of scheduled maintenance is a home kitchen gas oven.

Almost every oven users know that after some days of regular use of oven, one or more knobs of gas control get sticky. It needs a little force to operate the knob. In addition to this, flame of burner gets weak. This is due to the fact that traces of oil and food stuff keep on getting into the running clearance between the moving and stationary parts of knobs. This situation calls for immediate maintenance. This might not be immediately possible as the cooking might be going on. Other reasons may be that authorized maintenance technician may not be immediately available. Keeping all the above factors in mind, a monthly maintenance may be a regular action. Period of one month is just an example. It may be quite different for other plants and machinery.

Another significant example of scheduled maintenance is that of an aircraft.

There are thousands of components and assemblies which have got to be in perfect *health* as reliability of air worthiness of an air craft depends on it. All parts, components, assemblies are designed and made to great perfection.

Still natural wear cannot be avoided. What can be done is to replace parts or components which *lived* their safe working life.

Designers, manufacturers, quality experts and maintenance team decide and record as to what parts or components are to be replaced after how many working hours. There may be certain maintenance procedure where fasteners and joints are checked after specified period of air craft flights. Most of the times, parts or components are replaced, when still serviceable but near the end of *service life*. It can very well be imagined that if such components are allowed to function and if they fail during flights, then there may be emergency situations. For this reason, replacements are done at pre-determined time. This is called *scheduled maintenance*.

Breakdown Maintenance

There are instances, when an injection molding machine or an extruder stops working unexpectedly. Reason of stoppage may be minor such as blowing off of fuse or tripping of overload master circuit breaker (MCB). Major reason of stoppage of extruder may be breaking of extruder screw. In both the cases, it may be necessary to attend breakdown immediately or it may be differed for taking up in maintenance shift. It all depends on production schedules and demands. In case there is an urgency of completing production target, then maintenance may be taken up on *war footing*.

Maintenance team may have been working on some other maintenance jobs, when breakdown of extruder takes place. Since there is an urgency for restart of extruder, team may be called to attend to this breakdown first.

It is worth mentioning that breaking of extruder screw is a rare happening. In case it does happen then either broken screw may be repaired by making holes in broken parts and joining broken portion with the help of a steel insert and then both the pieces may be welded.

This arrangement is shown in Fig. 4.3.

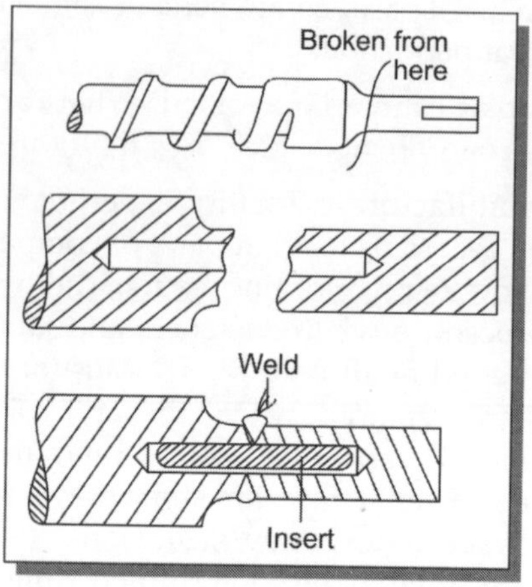

Fig. 4.3: Repair and broken extruder screw

Purpose of using insert is to keep both the pieces well aligned, when welding takes place.

Whole process of dismantling the extruder assembly, taking out the broken screw from cylinder, drilling of holes, making of mandrel (insert rod) and welding may take one to one and a half shift till production could be resumed.

In case a spare screw is available, then maintenance may be done in much less time. Repairing of broken extruder screw can then be done with normal working system. Repaired screw may again be fitted at some convenient occasion and new screw may be taken out and again kept as spare.

HYDRAULIC PRINCIPLES

Most of the plastics processing machines use hydraulic in addition to mechanical, electrical, electronics and pneumatic systems. From maintenance point of view, it is necessary to know about hydraulic principles and systems. Effective maintenance cannot be done fast unless concept of construction of hydraulic elements is clear.

Injection molding machines are mostly designed and made using hydraulic elements and circuits.

Engineering hydraulics deal with generation of hydraulic power, control, distribution and putting hydraulic power to work.

Principles of Hydraulics

Fluids is a term used for air, gases and liquids. Air and gases can be compressed in a confined space. Liquids cannot be compressed in a confined space, if force is applied. Liquids under pressure in a confined space exert same pressure at any point on the inside surface of confinement. This principle is illustrated with the help of Fig. 4.4.

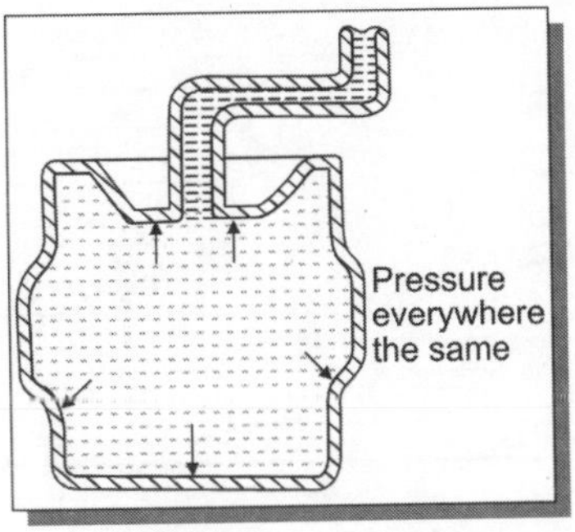

Fig. 4.4: Pressure distribution

Further *mechanical advantage* may be achieved with ease as compared to leverage effect. This point is illustrated with Fig. 4.5.

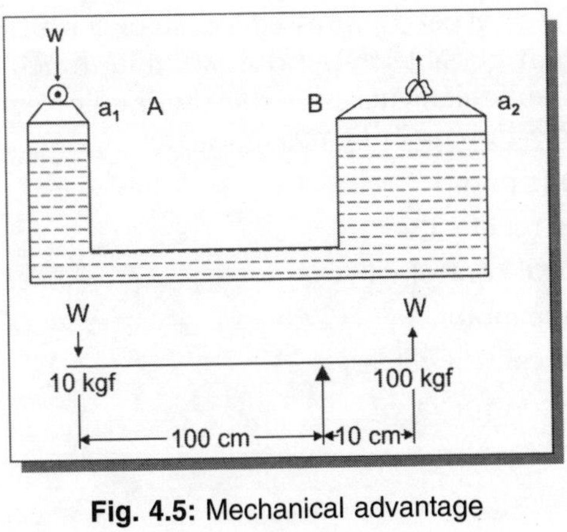

Fig. 4.5: Mechanical advantage

Pressure P_1 in smaller cylinder = w/a_1 kg/cm^2

Pressure on A = 10/10 =1 kg per sq cm

Same pressure would be exerted at B

Hence, available force would be

= pressure × area

= 1 × 100 sq cm

= 100 kg

Classical example of achieving *mechanical advantage* is a hydraulic jack for lifting wheel of a vehicle (Fig. 4.6).

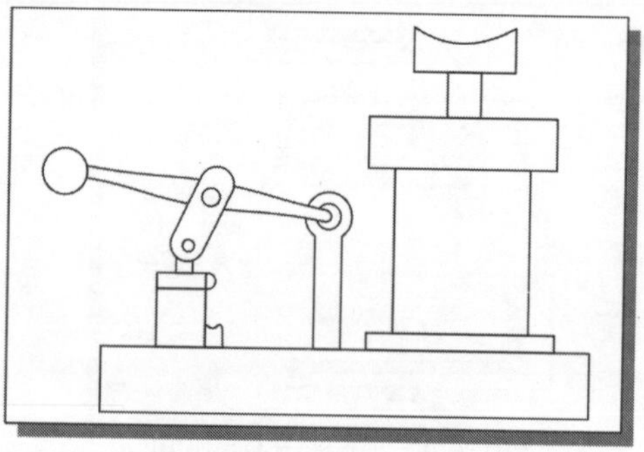

Fig. 4.6: Hydraulic jack

In hydraulic system of injection molding machine, hydraulic oil is the medium of working. It is worth mentioning that liquid in general does not get compressed like air or gases. This property of oil makes it possible to perform accurate functions through various hydraulic elements. Important hydraulic elements as used in plastics injection molding machines are:

- Hydraulic pump
- Pressure relief valve
- Directional control valve
- Hydraulic motor
- Pressure control valve
- Volume control valve
- Strainers
- Filter

- High pressure steel pipes
- High pressure flexible pipes (hose)
- Hose fittings.

Hydraulic Pumps

There are following types of pumps.

- Gear pump with external teeth (Fig. 4.7)

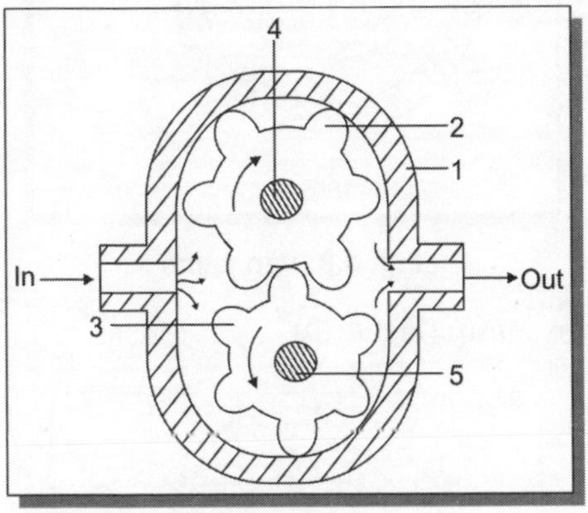

Fig. 4.7: External gear pump

- Gear pump with internal teeth (Fig. 4.8)

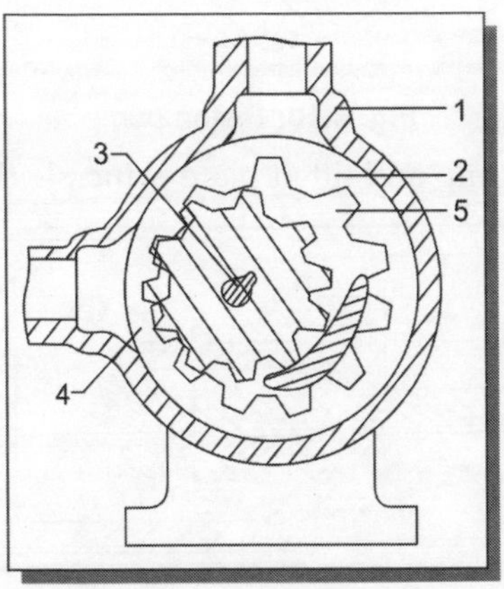

Fig. 4.8: Internal gear pump

- Fixed displacement van pump (Fig. 4.9)

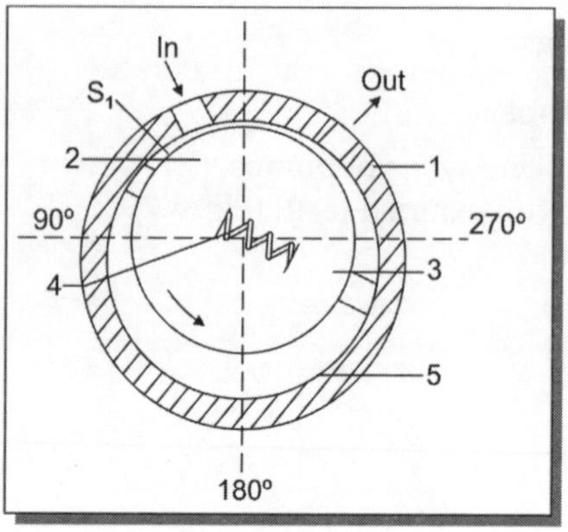

Fig. 4.9: Van pump

- Radial piston pump (Fig. 4.10)

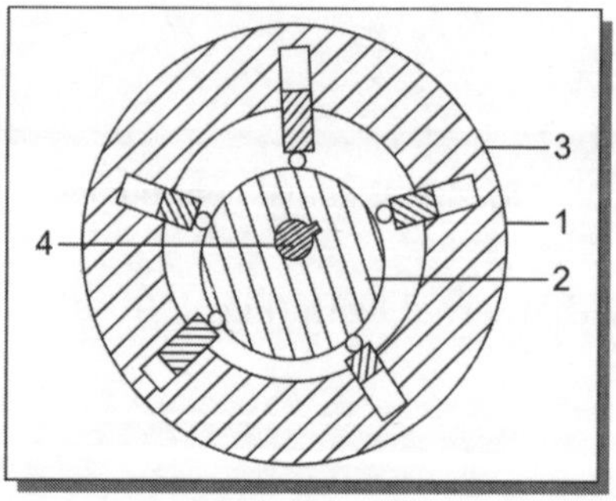

Fig. 4.10: Piston pump

- Axial piston pump and tilted plate principle (Fig. 4.11)

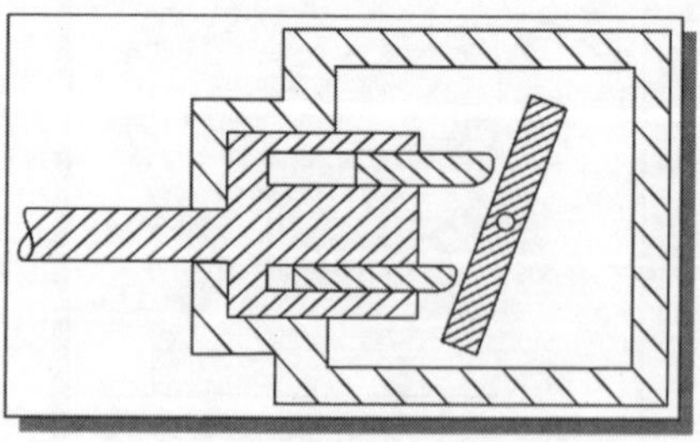

Fig. 4.11: Tilt plate pump

Referring to Fig. 4.7, typical basic construction of a external gear pump is shown.

Housing (1), gears (2) and (3) are so designed and assembled that there is a precision running clearance between gears and housing surfaces. Surfaces of gear teeth are so precisely machined, honed and polished that surfaces of teeth of two gears mesh very nicely. When gear (2) rotates in clockwise direction, gear (3) rotates in anti-clockwise direction. Oil carried by space between teeth of gear (2) and gear (3) ultimately reaches in outlet space. Oil cannot leak between the meshed gears, therefore, all the oil is forced through outlet of pump. Suppose outlet of pump is closed, then driving electric motor would get overloaded and get tripped. Working principle of internal gear pump (Fig. 4.8) is like that of external gear pump.

Working of fixed displacement van pump as shown in Fig. 4.9 is like this. (1) is the housing which has highly accurate and polished bore. A rotor (2) is much less in diameter. It is so assembled that its diameter almost touches bore of housing throughout its length. In fact, there is a clearance of 0.01 to 0.015 mm.

This clearance is enough for running of rotor without actual metal contact. At the same time, such a small gap does not allow oil to leak from high pressure outlet area to suction side. Rotor carry two precisely sliding vanes (3). In between the two vanes, a spring is shown which keep the vanes edges pressed against bore surface (3). In actual practice, there are a number of small vanes without any springs. High centrifugal force tends to throw vanes out. Consequently, edges of vanes always remain in firm contact with bore surface when pump is running. Let rotor be rotating in anti-clockwise direction then space S_1 keeps on increasing in volume thus sucking in from suction port. When other vane comes after crossing seal line, it encounters with already filled space. Further rotation of rotor makes this vane to force the oil out of outlet port.

Radial Piston Pumps

Typical basic construction of a radial piston pump is shown in Fig. 4.10. (1) is the housing of pump having suitable holes and channels for oil to flow in and out. Rotor (2) is an eccentric cam which keep on pushing and pulling the pistons in their cylinder holes. Oil is sucked in, when pistons are moving towards the centre and force the oil out, when pistons are moving towards outer periphery.

Axial Piston Pump and Tilted Plate Principle

Typical basic construction of a axial piston pump with tiltable plate is shown in Fig. 4.11. In this pump, it is possible to vary output (volume) of pump.

It is used in many applications including big size (capacity) injection molding machine. These types of pumps are also used for other industrial and field operation machineries.

Almost all hydraulic pumps have a service life in terms of operational hours. After expiry of service life, efficiency of pumps starts deteriorating. Mainly two things happen. Maximum value of output pressure decreases and delivery at a specific pressure starts decreasing. This often happens due to following reasons:

- Oil seals become worn out
- Clearance between moving and stationary parts increases due to wear.

In many cases, performances of pump can be restored by changing all rubber seals and O rings. In case rotor, vanes or stator rings are worn out then it is necessary to replace kit which contains new rotor with vanes and stator with a set of seals and O rings.

Pressure Relief Valve

Pressure relief valve is a device which allows hydraulic oil to flow out to the tank in case pressure rises above its specified limit. If a relief valve is not provided and flow of oil from pumps get restricted or completely stopped, then there is a great likelihood of pump or its drive getting damaged (Fig. 4.12).

There are a variety of relief valve, such as:

- Simple relief valve
- Pressure override
- Compound relief valve
- Balanced piston
- *R* type valve.

Basic construction of a typical simple relief valve is shown in Fig. 4.13.

(1) is a cap having a seat to block the nozzle from which oil can enter. Cap is kept pressed against nozzle with the help of a compression spring (3). Compressive force of spring can be increased by screwing down knob (4). Cap gets lifted, when oil

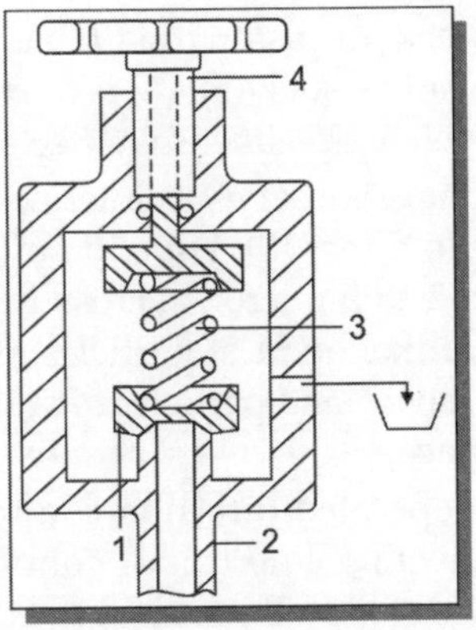

Fig. 4.12/4.13: Relief valve

pressure in nozzle (2) increases beyond set limit. Most single stage vane pumps are capable of delivering hydraulic oil at up to 2000 psi. Generally, in practice, working pressure is kept below the maximum.

Accordingly, relief valve is set to open at, say 1700 psi.

Directional Control Valve

Purpose of directional control valves is to direct pressurized oil from pump to a particular location in the hydraulic circuit. Flow can be changed to other location. Figure 4.14 shows a typical hydraulic circuit to activate a jack.

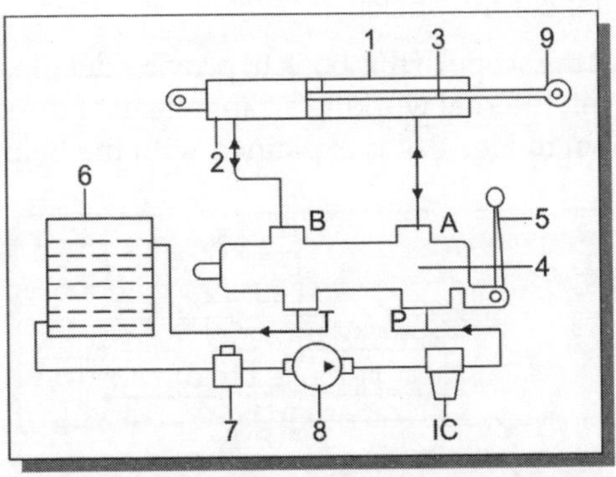

Fig. 4.14: Directional control valve and jack connection

(1) is a hydraulic jack. Its piston rod (9) has to be moved IN and OUT by a manual control stick (5). (6) is a hydraulic oil reservoir from the bottom level, a pipe line is connected to strainer (7).

From strainer, connection goes to inlet of pump (8). Outlet of pump is connected to filter (10). Function of filter is not to allow particles more than 2 to 5 microns across to pass. If filter is not used, then there is a likelihood of particles of 10 microns pass to directional control valve. Solid particles of around 10 microns may damage sliding surface.

(4) is a manually operated direction control valve. It has four ports. P is for connecting line (pipe) coming from filter. A is connected to cap side of jack. B port is connected to block side (2) of jack. Port T is connected to tank (6). When stick (5) is in neutral position, hydraulic oil entering into directional control valve passes to tank from port T. There is no high pressure oil passing either to A or B. When control stick (5) is pulled towards right, oil flow shifts to port B, hence piston rod moves out. Any oil on the other side of piston goes to tank through port A. When control stick is shifted to left, direction of flow of oil gets reversed. Consequently, piston rod moves in.

There are a variety of directional control valves having complex functional characteristics. Most of the directional control valves are designed and built to operate by following means:

- By electric solenoid
- By pneumatic arrangement
- By low pressure hydraulic circuitry
- By manual control sticks.

It is beyond the scope of this book to provide detailed description of all. However, internal working arrangement of direction control valve as shown in Fig. 4.14 is explained with the help of Fig. 4.15.

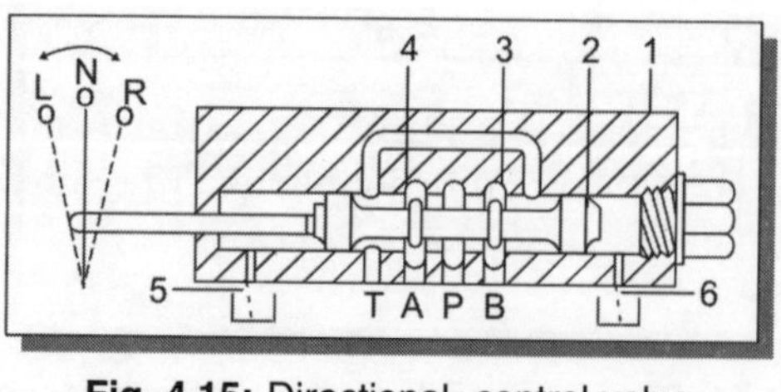

Fig. 4.15: Directional control valve

(1) is body of directional control valve. It has highly précise honed bore in which spool (2) moves. (3) and (4) are lands of spool. There are four ports:

P – Line from outlet of pump via filter is connected.
A – Line from one port of jack is connected.
B – Line from other port of jack is connected.
T – Line from tank is connected.

Spool is shown in neutral position, where flow of oil from pump directly goes to tank.

When spool is shifted towards right, oil flows from port P to B and at the same time oil from jack return to tank via ports A and T. If spool is shifted towards left, oil flows from port P to A and at the same time oil from jack returns to tank via ports B and T.

Manufacturers of directional control valves produce a variety of valve to suit varied requirements. For example, directional control valve shown in Fig. 4.15 may have other ports configuration so that in neutral position port P for entry of oil from pump is blocked so that oil under high pressure is utilized for operating other parts of circuit, say another jack or a hydraulic motor.

Availability of various types may be found out from technical literature issued by manufacturers like Vickers India. In fact a number of reputed manufacturers of directional control valve may be located from internet search sites like Google or Yahoo.

Hydraulic Motors

Basically there are three types of hydraulic motors:

1. Gear
2. Vane
3. Piston

All above types of motors have a variety of designs and functions.

Pressure Control Valve

Purpose of this valve is to control pressure for a particular section of hydraulic circuit. Basic functioning of a pressure control valve is explained with the help of Fig. 4.16.

(1) and (2) are two ports, one for incoming and other for outgoing hydraulic line. Outgoing hydraulic line pressure can be reduced by this valve. (3) is the space through which hydraulic oil passes

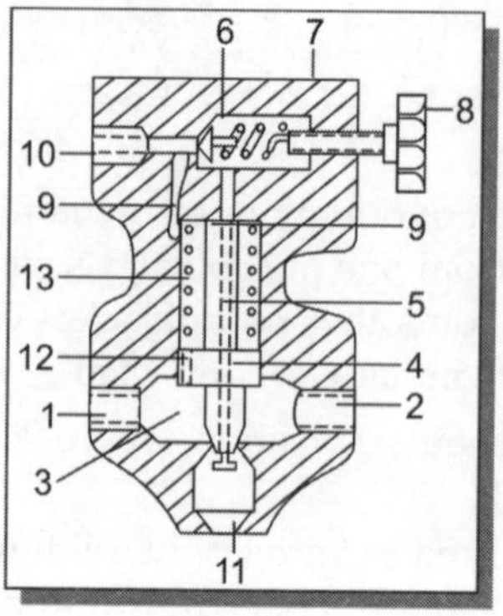

Fig. 4.16: Pressure control valve

and is under pressure. (4) is a piston which can precisely slide in bore. Since there is a small hole (12), oil pressure on both sides of piston is same. Piston has hollow tronion (piston rod) in which a tension spring (5) keeps on pulling the piston up. Complete length of piston, including rod is hollow from which oil can pass from upper chamber (7) to tank through opening (11). Upper chamber has a spring loaded pin (6). Spring force can be adjusted by a knob (8). Chambers (3), (13) and channels 9 and 10 remain filled up with high pressure oil. Chamber (7) and hollow of piston remain filled up with low pressure oil. Knob (8) is used to adjust spring pressure on cone pin (6). This keeps the channel (10) blocked with a certain force adjusted by knob. In case pressure in chamber (10) rises and cone valve opens, this creates a connection to low pressure.

Consequently, piston (4) rises a little to provide passage of oil from chamber (3). As soon as pressure of oil drops, cone pin closes the passage (10) again. In this way, self and continuous maintenance of pressure is achieved.

Volume Control Valve

Speed of an actuator depends on the quantity of oil flows in per unit of time. Flow of oil can also be controlled by a variable output pump. But in most cases, with fixed output pump, a volume control valve is used.

A schematic Fig. 4.17 of a non-compensating flow valve is given here. Its function is explained as follows.

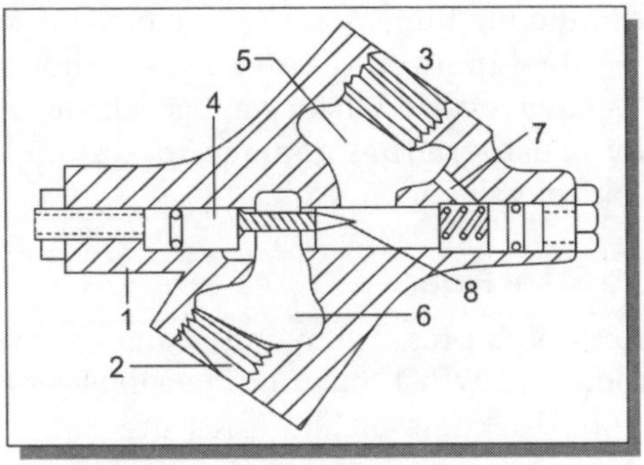

Fig. 4.17: Volume control valve

(1) is the body of non-compensating flow valve. There are two ports (2) and (3). (4) is sliding piston. Left hand face of piston is stopped by an adjustment screw. Other face of piston is kept pressed by a compression spring. There are two cavities, (5) and (6). Cavity (5) is connected with space where the spring is connected by a small passage (7). If port (2) is used as inflow, then hydraulic oil will push the piston against springs. Consequently, cavities (5) and (6) will get connected and there would be free flow of oil. In case port (3) is a input passage then the only passage for oil to pass from cavities (5) and (6) through triangular cuts on diameter towards cavity (6). Volume of oil flowing would depends on as to how much length of cuts (8) are exposed to cavity (5). This depends on the adjustment of screw (9).

Strainers

Strainers are course filters which are generally fitted on the suction line of pump. Strainers are so fitted that they remain submerged in oil of the tank. Strainers may be said to be course filters. Function of strainers is to stop contaminant particles flowing to pumps. Sizes of particles are generally denoted in microns. Strainers generally have fine wire mesh of 100 mesh. They protect pump from particles above 145 microns size.

Filters

Filters are normally fitted beyond pump outlet. These are made from many materials. Filters are designed and made to arrest particles more than 10 microns or even more fine particles.

Clean hydraulic medium is an issue which constantly being examined by designers and filtering technology experts. Investigations have revealed that even particles as small as 2 to 4 microns may cause gradual damage to system units, say a directional control valve.

High Pressure Steel Pipes

Pipes as used in high pressure hydraulic lines are seamless and are made from steel which has high tensile strength and rust resistance. Wall thickness of high pressure pipes depends on diameter of pipes and tensile strength of steel.

Hoses

These are flexible *pipes* made to withstand high pressure. They are fabricated by using layers of synthetic rubber and braded fabric or wire. These hoses are used in between two units which have relative motion. Sliding injection unit of a injection molding machine is an example where hydraulic jack ports are connected to steel piping by means of hoses.

Joints

Most of the joints in hydraulic system are based on thread. Male and female components are tightened along with some type of seals such as synthetic rubber, plastic and copper washers of various shapes and dimensions.

At many places, metal cones are used to create a seal at the joint. Cones are tapered component of softer metal. These are placed in between joining surface and then tightened by threaded components. Few joints are shown in Fig. 4.18.

ELECTRICAL CONTROLS

All plastics processing machines are driven and controlled by electrical power at various voltages. Pumps, extruder screws, calendaring rollers, vacuum pumps of vacuum forming machine are driven by electric motor which normally requires 3-phase, 50 Hz, 440 volt AC. Variable speed motors are also used on places such as granulator rotor cutter. These motors are generally run by DC voltages where step less speed variation is possible with ease. Variable speed AC motors are also available.

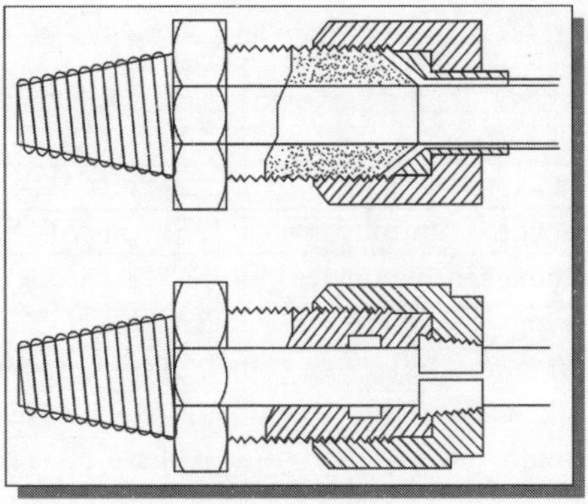

Fig. 4.18: Joints

In hydraulically operated machines, a number of solenoid are generally operated by 220 volt, 50 Hz AC supply.

Mold locking, injection nozzle forward and injection, etc. are achieved by energizing directional control valves in a sequence which is generated by various timers, sensors and electronic circuitry. Sensor and electronic circuit are energized by low voltage DC supply.

Now DC voltages, are obtained by *power supplies* which are incorporated in electrical system of machine.

Voltage like +5, –5 and +12, –12 are tapped out from *power supply*.

Table 4.2 gives an idea of assigned voltages to various electrical elements (components).

Electrical cum Electronic Control of Injection Molding Machine

Electrical cum electronic circuits of plastic processing machinery are designed in such a manner that desired sequential functions are achieved manually, semi-automatic and fully automatic. In this portion of chapter, it is attempted to explain how the wiring for an injection molding machine is done to achieve inching, semi-automatic and automatic operation. Please note that given wirings diagram is just to explain. Actual wirings in machines are complicated and need wiring diagram from machine manufacturer to understand real wirings.

Explanation of wiring diagram is done for a typical sequence of operation of an injection molding machine. It is given as follows.

Table 4.2: Operating voltages for some electrical gadgets

S. no.	*Electrical components*	*Generally assigned voltage*
1.	Audio alarm	220 volts AC
2.	Contactors	220 volts AC
3.	Directional control valve	220 volts AC
4.	Earthing sensing device	5 volts DC
5.	Heating	220 volts AC
6.	Light lamps, LEDs	220 volts, 3 volts DC
7.	Low voltage sensors	12–18 volts DC
8.	Motors	3 phase, 50 Hz, 440 V AC
9.	Overload protection	220 volts and 6 V DC
10.	Pressure sensors	6 V DC
11.	Solenoids	220 volts AC
12.	Switches	—
13.	Single phase protector	220 volts AC and 6 volts DC
14.	Timers	6– 18 volts DC

Machine is switched on and started:

- Moving plates move forward.
- If there is no hindrance in between two halves of mold, only then full locking force is generated.
- When mold completely gets locked, then plasticizing unit moves towards the back of mold.
- Nozzle of plasticizing cylinder sits on feed head of mold with a predetermined force.
- As soon as nozzle sits on feed head, injection screw plunger moves forward to inject plastic into mold.
- After a laps of few seconds, plasticizing screw starts rotating, thus pushing back injection screw plunger. By this time, there is no full injection pressure inside the cylinder. Only a small pressure remains. Oil in the cylinder at the back side of piston is gradually pushed by rotating and linear moving screw. This action provides *back pressure* to plastics.
- When screw stops rotating, injection unit moves back.
- After the unit is completely moved back.
- Needed extra time is provided for mold to remain close for allowing molding to cool down to required stiffness for ejection.

- After elapse of mold stay time, mold opens completely.
- Hydraulically or pneumatically operated ejector rods operate to eject molding (components plus runners and sprues).
- Now one semi-automatic cycle is completed.
- For activating next cycle, start button has to be pushed again.

An electrical cum electronic circuit diagram for above mentioned sequence would be quite complicated. Various machine manufacturers generally provide circuit diagram for their machines. This helps maintenance team of machine user to locate faults and carry out repairs. Further, specifications and code number of electrical and electronic parts and units are also provide for ordering spare parts.

However, electrical circuit diagram for a simple all *electro-mechanical* machine is given here to explain wiring of various electrical and electronic parts and achievements of sequential operation cycle of molding.

Fig. 4.19 shows a typical all electro-mechanical machine for explaining sequence of operation. Electrical circuit (Fig. 4.20) would then be explained for machine shown in Fig. 4.19.

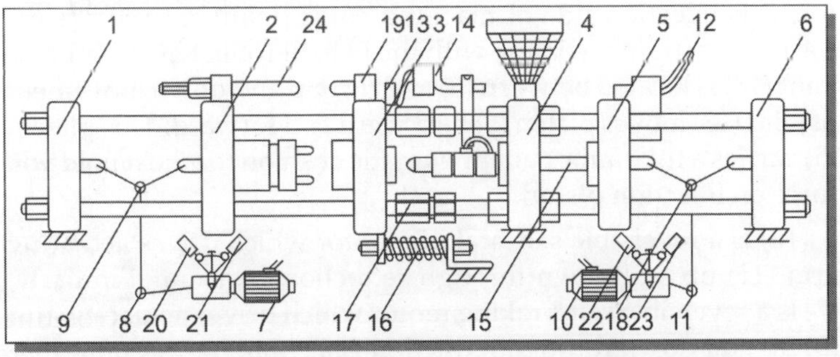

Fig. 4.19: Electro-mechanical injection molding machine

There are six plates. (1) is fixed platen towards locking side of machine. (2) is the platen which is movable on tie bars and carries ejection side of mold *half*. (3) is a floating platen which carries injection side of *half* mold. This platen is loaded with two systematically placed compression springs on rods. These springs and rods keep floating platen (3) about 30 mm away from four robust supporting nuts on four tie bars. (4) is a fixed platen which

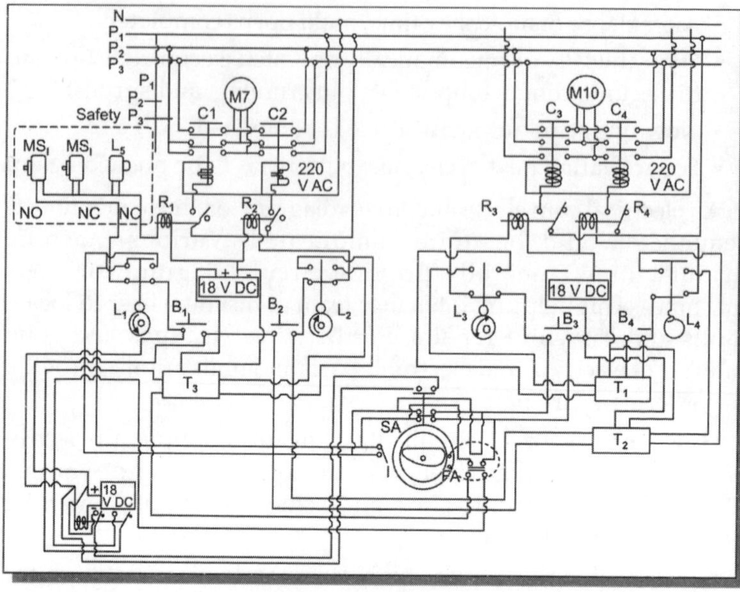

Fig. 4.20: Electrical circuit

carries plasticizing unit and hopper. Platen (5) is a moving platen towards injection side of machine. (6) is finally a fixed platen. Tie bars are tightened here with the help of robust nuts. Knuckle joint (18) is located between (5) and (6). Up and down movement of joint (18) moves platen (5) backward and forward. This platen, (5) carries a hydraulic cum pneumatic cushion to adjust injection force on injection piston.

(10) is a reversible self-braking motor which drives actuating arm (11) up and down through reduction gear unit. Similarly, (7) is a reversible self-braking motor which drives actuating arm (8) up and down through reduction gear unit.

There are limit switches (20) and (21) on locking side drive unit. These limit switches are actuated by adjustable cam segments fitted on shaft of actuating arm (8). Similarly, two limit switches are there on injection side drive unit.

On platen (3), a micro-switch (19) is mounted and on platen (2) an adjustable rod is fitted which just presses micro-switch (19) when mold is completely closed against force of two compression springs (16). There is another micro-switch (14) which gets just actuated, if floating platen (3) completely sits against robust nuts on four tie rods.

Complete locking would take place, when both micro-switches are pressed. In case there is some obstruction between two halves of mold, then micro-switch (19) would not operate even if platen (3) reaches to the faces of all the four nuts (17). Since micro-switch (19) has not operated, motor (7) of drive unit will stop and simultaneous braking action takes place and further sequential operation are aborted. This saves the mold from getting damaged.

In case there is no obstruction and mold closes normally then both micro-switches get operated. Hence, sequential operation gets completed. There are following three options for machine operations and it is manually selected by a selector switch on control panel.

Inching

In this mode, mold locking, opening can be done by momentarily pressing the inching button as many times as required. This mode is generally used during mold setting or while examining if mold closing and opening is correct.

Similarly, injection piston can also be inched forward and backward.

Semi-automatic

In this selected mode, only one molding cycle takes place which ends with moving platen (2) at complete back position where ejection of molding has also taken place.

Fully Automatic

Selection of this mode of operation makes the machine to repeat molding cycle automatically after providing some stay time (ejection time) between two cycles.

All the above described sequential operations are affected by the use of following electrical elements:

- Selector switch
- Relays
- Contactors
- Limit switches
- Micro-switches

- Push buttons
- Timers
- Motors
- Main switch
- Over load protection
- Connectors
- Wiring
- Optical indicator
- Warning light
- Flap switch
- Temperature indicator cum switches
- Power supply.

A typical wiring diagram is given here to explain as to how sequential operation takes place at any of three setting options, viz inching, semi-automatic and automatic. First of all, few points are clarified as under.

- At inching position of selector switch, there are two contact points which always remain open, even if knob is rotated to other position.
- At semi-automatic (SA) position, there are three pairs of contact points which remain open, when selector knob is at *I*. All the three pairs of contact points get closed, when knob is brought at SA position.
- Again at FA (fully automatic) position, there are three pairs of contact points which remain open, when knob is not at this position. All the three pairs of contact points get closed as soon as selector switch knob is brought at fully automatic (FA) position.
- Direction of rotation of 3-phase motor depends on internal connection of windings. Two sets of wires come out of motor. Each set has 3-wires. If one set of wires is connected to 3-phase supply then motor will rotate in a direction, say anticlockwise. On removing first set of wires and connecting other set would rotate motor in clockwise direction.
- Both the motors are self-braking. As soon as supply of current is disrupted, instantaneous braking takes place. Consequently, movement of platen stops instantly.
- Two contactors are used for each motor (locking side M7 and injection side M10) for change over direction of rotation.

- 220 V AC supply is given to contactor coils through relays which are operated by 18 volts DC supply.
- Two power supply units are used. One for locking side and other for injection side.
- There are two limit switches L_1 and L_2 (circuit) on locking side and two on injection side. These are actuated by means of cams mounted on shafts which move toggle joint up and down. Position of these cams can be shifted with respect to shaft on which they are mounted.

There are two safety micro-switches MS_1 and MS_2. There is a limit switch LS for safety door.

Operation of Circuit

Let machine is totally off. Mold is completely in open position. Injection unit is completely in out position (towards right hand) operation in inching.

- Put the main switch of machine ON.
- Bring the selector switch on inching position.
- Press button B_1 momentarily. It allows 18 volts DC current to pass through relay R_1. Relay operates and allows 220 volts AC to pass through coil of contractor C_1.
- Motor M7 rotates in anti-clockwise direction as long as button B_1 is kept press momentarily.
- Platen moves forward, MS_1 gets operated and points get closed, if there is no obstruction between mold halves. Movement of platen P_2 (if inched) reaches its extreme on support nuts. At this time, MS_2 opens. But in effect circuit here remain closed as MS_1 is closed and is parallel with MS_2. Safety door is closed and hence limit switch LS has closed point.
- If there has been an obstruction, then this portion of circuit would remain open and thus current will not pass through relay R_1. Motor will instantly stop no matter how many times switch S_1 is pressed.
- Once the mold is satisfactorily closed and injection nozzle is properly sitting on feed head then press button B_3 momentarily or continuously (as desired) to inject plasticized plastic inside the mold. Injection platen (5) reaches its limit and motor stops as passage of current to relay (3) coil gets disrupted by limit switch L_3.

- By pressing B_4, platen (5) can be inched back to extreme right position. In case of pressing B_4, direction of rotation of motor M10 reverses through relay R_4, which operates contactor C_4.
- Now operator may wait for few seconds or minutes before pressing button B_2 to open mold. Pressing of B_2 reverses the rotation of motor through relay R_2. Platen (2) would move to its extreme left, if B_2 is kept pressed. On reaching extreme position, L_2 would get operated by cam and flow of current to relay R_2 would get disrupted. Hence cycle on inching is completed.

Semi-automatic Operation

- Bring knob of selector switch on SA position. There are three pairs of contact which get closed.
- Press button B_1, 18 volts DC current passes through relay R_1 which in turns activates contactors C_1. 3-phase 440 volts pass through motor which instantly starts. In SA mode, even releasing of button B_1 will not stop the motor as circuit is latched through latching circuit LC which was inactive during inching selection.
- On completion of full locking, limit switch L_1 closes contact of B_3 without actually being pressed.
- Relay R_3 is energized which in turn connect coil of contactor C_3 to 220 volts AC. Motor M10 gets 3-phase 440 volts supply and starts to move injection platen (5) to its full limit, when limit switch L_3 operates as cams raised portion finishes and roller of limit switch goes down thus actuating L_3.
- For providing few seconds to keep injection piston in forward position (plastic being kept pressed inside plasticizing cylinder).
- L_3 in its low position triggers timer T_1. After set time of few seconds, T_1 provides passage of 18 volts DC to relay R_4 which gets activated.
- Contactor C_4 energizes and motor M10 starts in opposite direction till limit switch L_4 reaches its limit (injection platen is now at its right side position).
- Change of position of roller of limit switch L_4 triggers timer T_2 which provides some seconds for mold to remain closed.
- After elapse of mold stay times from T_2, it triggers relay R_2. In turn it activates contactor C_2 to start motor M7 in reverse direction.

- Mold platen (2) travels to extreme left till limit switch operates and motor M7 stops instantly.

Next cycle will not begin unless button B_1 is pressed once again.

Now to start fully automatic mode, selector switch knob will have to be turned to position FA (fully automatic). It will bring a third timer T_3 into circuit which provides time for ejection of component, runner, etc.

- To initiate fully automatic cycle, press button B_1.
- On completion of full once cycle, position of switch L_2 changes and timer T_3 is triggered. After completion of elapse time (ejection stay time), timer T_3 provides a close contact to B_1 without B_1 being pressed.
- Hence next molding cycle starts automatically.
- In case during mold closing, if there is any obstruction due to even mold malfunctioning, machine would stop.
- There may be an emergency big button in red color to stop machine in emergency.
- Just push and machine stops no matter at what stage the cycle machine is. This red emergency button disrupts 3-phase 440 volts supply.

ELECTRONICS AND MICROPROCESSOR BASED COMPONENTS AND EQUIPMENT

It is already seen in above sub-heading that a number of electrical and electronic components are used to built up a logic circuit to create desired sequential operation of machine. This methodology is quite time tested. But needs much space to assemble all the components and gadgets. Further, making of such circuit needs a lot of labor which increase the cost of machine. Development of new technologies took place over a period of time. Improved design and construction of machine controls came into exhistance.

There are many types of plastic processing machineries. Each machine may have different operations to perform or may have different sequence of operation. Therefore, electrical and electronic components and wirings would differ from machine to machine.

Jobs of machine design and building are made easier by using following high tech controllers.

PID — proportional integral derivative
PLC — programmable logic controller

PID

PID is acronym of proportional integral and derivative controllers. This controller works on close loop technology. PID is a combination of three terms, proportional, integral and derivative. Word *term* is used to convey ideas of totality proportional, integral and derivative. All the three terms in a controller are called PID.

Example of temperature control of plasticizing cylinder is taken to explain PID. Temperature of cylinder is sensed by a thermocouple.

Suppose target temperature of plasticizing cylinder is 210°C which is indicated by a digital temperature cum controller. As soon as temperature reaches set point and temperature controller sends a signal to contactor of heaters.

Temperature overshoot by say 2°C and then effect of supply cut start dropping temperature till restart of heating is triggered. This effect does not allow to control of temperature within more elapse limit, say ±0.05°. Heating is much better controlled by adjustment of temperature proportional to variation which practically occurs. This is one function of controller. Other function is integration of many occurrence of variation in temperature. This takes care of variation due to a number of causes, hence provide a better control over the temperature. Third term derivative *calculates* (as PID may be considered a minicomputer) best control level and provides output to trigger heating circuit contactor. Similarly, PID may be used for accurately controlling hydraulic pressure and volume or speed of an actuator.

Individually, a proportional controller may not provide SSE (system specification extension) performance required for the system.

An integral controller may provide SSE performance while slowing down system. If derivative term is used, then required performance of system may be achieved and at the same time speed of system may remain normal. In a PID, three controllers act in combination, which are proportional, an integral and a derivative.

PLC

PLC is acronym of programmable logic control. It is used to control operations and sequential function of various plants and machinery. PLC is like a mini-computer which receives input signals, process these signals through control logic and then produces output signals which are utilized to turning output ON and OFF depending on state of input signals.

PLC can *read* and *write* various types of signals which may be discrete (digital) and analog. Digital signal means on and off, 1 or 0, high or low.

Digital signal, for example, can activate a motor to start or stop. Analog signals are *flowing* signals which may be between two limits. Analog signals may have different values of pressure, temperature, speed, etc.

Latest PLCs are built around hardware components which are basic for many system. These are processors which consist of CPU, RAM and interface capabilities.

- IN and OUT outputs
- Power supplies
- Circuit loaded card and PCBs
- Communication interface to interact with operator or programmer, data monitoring system and remote input and output
- Programming facility and access to processor.

Referring to wiring diagram of all electro-mechanical injection molding machine, a number of relays, micro-switches, limit switches, selector switch and push button are used to create the logic for sequential operation of the machine.

Wiring of these components is done in such a fashion that desired sequence of operations is achieved.

Use of PLC may do away with most of conventional wiring. Further, desired logic of sequential operations may be fed to memory of PLC by means of a software CD. Now the input to PLC by operator would be to press button or select on display screen a mode of operation, inch, semiautomatic or automatic and then to press start or touch start on display screen. By doing this, PLC would deliver output signals to various components.

Change of mold for another plastic component may need different setting of time, temperature speed and even sequence for a hydraulic machine. These changes may be automatically affected merely by feeding data through software CD. Use of CD may not be needed, if setting parameter for various molds are already stored in memory of PLC.

Sometimes, it becomes necessary to fine tune parameter of a particular mold. This can easily be done on PLC.

Capabilities of PLC are getting improved day by day to precisely handle much more complex operation such as continuously monitoring sheet thickness in extrusion process and to automatically keep on adjusting die opening to maintain target thickness under tolerance limit. PLC can also monitor Parison profile of a blow molding machine. Suitably installed sensors provide input signals to PLC which in turn sends out signals to adjust speed of extrusion, die opening profile, etc.

MANAGEMENT OF PLANT AND MACHINERY MAINTENANCE

Ultimate objective of a production organization is to produce maximum quantities of product with maximum productivity. This means that minimum numbers of worker and operators are deployed to produce certain quantity of product. If number of persons deployed increases for the same output then the quantity of product, produce per person can be calculated and it would be found that output per person has reduced. Hence it may be said that productivity has gone down. In case number of person deployed is reduced and production output remains the same then it may be said that productivity has increased.

Similarly for producing specified quantity of product, some machine hours are needed. Let us take example of a PET bottle producing plants. In this typical example, plant operates on two shift basis. There are five-blow molding machine and two-injection molding machine.

Each machine is allowed 15 minutes for daily routine clean up and check up. Hence productive time for each machine on two shift basis would be 15 hours per 24 hours. Hence productive time for five machines would be 75 hours. If machine hours for any reason gets reduced then productivity per machine goes down. So to avoid reduction in machine productivity, machine should be kept production worthy at all times. This may be possible, if maintenance of machines and auxiliary equipments is done effectively and within minimum possible time. Therefore, maintenance of machine need to be effective, fast and with just sufficient manpower. For all these, efficient management of maintenance is needed.

Management of maintenance needs following inputs:

- Availability of maintenance manual for each type of machine
- Trained and dedicated fitters, electricians
- Electronic technicians
- Availability of spare parts
- Working tools and jigs
- Recorded maintenance history of each and every machine
- Capable leader of maintenance team
- Coordinator maintenance and production shop
- Deployment of external maintenance agencies for specific maintenance work or in case of emergency
- Analyzer to analyze maintenance effectiveness trends and cost of maintenance as percentage of total production cost of products over a period of time, say a month or year
- Attractive service terms and condition
- A methodology in place to establish if breakdown are due to negligence on the part of operator or maintenance workman
- Well defined job description of members of maintenance team
- Attitudes of leader of maintenance team and that of coordinator maintenance and production team leaders
- To ensure that members of maintenance team are not harassed, team leader should keep the moral of team high
- Periodical assessment of performance of each and every member of maintenance team
- Some suitable incentives scheme may be in place.

Manuals

Manufacturers of plastics processing machinery usually provide *operating manual* for operating plant and machinery. In addition to this, they may also supply maintenance manual. These manuals provide sufficient guidance for carrying out maintenance. This does not include maintenance of bought out items such as pressure relief valve, directional control valve. Maintenance manuals for such items may be obtained from machine manufacturers who may have obtained manuals from manufacturers of relief valves, directional control valves, motors, etc. Users may also directly write to manufacturers of these items. Companies of good reputation may supply maintenance manual directly to user of plastics processing machine in which these machine elements are fitted.

Trained and Dedicated Maintenance Man Power

Person selected and employed for machine maintenance work may be fresher from ITIs or having few years of experience in the field of plants and machinery maintenance. Still, it may be necessary to impart training to make newcomers to become familiarized with new environment. They are also explained function, construction and maintenance needs for various machines.

Fresher may need more time and attention to become capable of assisting senior maintenance colleague. With passing times these new comers would become experienced to do work independently.

As for as dedication is concern, it is a personal quality of an individual. It cannot be infused, but it can be improved by motivation.

Electronic Technicians

In case system element such as sensor or *power supply* malfunction then an electronic technician is needed to locate fault. Electronic faults rarely occur, hence an electronic technician may be enough to handle work from plant, having number of machines.

Electronic gadgets like timers and PLCs generally contain edge connecting cards which may easily be replaced in case of malfunctioning. Defective cards may be sent to card repair specialist for repair. After repair, electronic technician may try out. If found OK, it may be left fitted on edge connectors. Taken out new card may again be stored as a *stand by* spares.

There may be a number of machines in a production shop which are to be kept production worthy. This needs urgent or periodic maintenance for which spare parts may be required. There may be a large number of spares which might be needed. Some of spare parts are frequently needed and some rarely such as an extruder screw or worm and worm wheel of reduction gear box. These machine parts are generally costly and rarely stored, if instant availability is possible. In case location of works is such that spare parts are not available and would require few days till spare parts is made available. Keeping a production machine non-functional would mean quite substantial loss of production. Therefore, maintenance management prefers to invest in rarely needed spares rather than incurring losses, due to production stoppage.

In maintenance work, specialized tools are also needed together with conventional tools. Specialized tools may include pullers, hydraulic jacks, chain pulley block, light duty cranes which move on floor with the help of rollers, heating blower, etc. Maintenance team member must very well know how to use specialized tools correctly.

Experienced workers generally know as to how these tools must be used safely and correctly. New members of maintenance team should be trained.

History Recording

In many production organizations, there is a system in place. In-charge production shop sends a requisition slip to maintenance department for carry out maintenance of a particular machine. A typical requisition slip is reproduced here.

Requisition Slip Date

From: In-charge molding section To: In-charge maintenance

Machine: Injection molding No: Bay 1/3

Defect: Full injection pressure is not developing

Vibration and sound, when injection pressure is applied.

Production done since last maintenance : Approximately 500 K

Maintenance required : Normal, urgent and emergency

Remark of maintenance in-charge: Would be taken up in maintenance

Shift 6 PM

Date: Time: Signature I/C maintenance:

Note: Original sent back to I/C production

Copy to be kept by I/C maintenance

Above information are punched into computer system installed in office/cabin of I/C maintenance. After maintenance is carried out, a report is briefly written by maintenance man and submitted to in-charge. This report may be just on a blank paper and it may be like this.

Machine no: B1/1, injection molding machine

- Oil seals of pump replaced
- Seat of pressure control valve

Re-lapped and polished, machine is now functionally OK

Date:
Time: Name of workman:

Maintenance data loaded in computer from time to time is analyzed by a maintenance software loaded on computer. In-charge maintenance may see, for example, which machine and its component is giving frequent trouble. How many times the defect in question is developed during last 6 months. This type of analysis may be seen for other similar machines also. It may be revealed that a particular machine is giving frequent trouble. Once it is known, a critical investigation may be carried out to find cause of frequent repetition of defect. It may be necessary to replace hydraulic pump by a new one.

As already mentioned, capable leadership is required to keep plant and machinery well maintained. This also needs good coordination between both in-charges to decide line of action best suited to the circumstances prevailing at a particular time.

Suppose bearing of calendaring rolls is worm out. Consequently, setting and maintenance of sheet thickness has become difficult. It, therefore, needs complete dismantling, repairing, replacement and reassembling. It is going to be a time consuming work and needs uninterrupted attention. Strength of present maintenance team cannot be deputed on this job as other day-to-day maintenance may suffer. For this reason, outside maintenance agencies may be engaged to carry out repairs on shop floor as machine cannot to shifted to other locations.

In almost all the organizations there are accounts and audit section. One of these departments or sections is assigned with the work of working out production cost on monthly and yearly basis. One of the fixed and variable cost factors is expenses on maintenance of plant and machinery.

Fixed maintenance cost includes:

- Salaries of team members
- Cost of shed and space allotted to maintenance section
- Depreciation cost of maintenance tools, jigs and repair machinery
- Other fixed cost factors which are related to over all plant operation

Variable maintenance cost may include:

- Value of spares used during specified period of time, say, a month or year

- Consumable items like oils, grease, cleaning cloth, washing liquid like kerosene, washing soups, etc.

Payment of bills from outside agencies is another variable expenses. Amount of bill depends on nature and extent of work carried out.

Many a times dispute crops up between in-charge of production shop and maintenance. When in-charge maintenance receives a requisition slip from production shop, he finds that repair or rectification of this particular defect was recently repaired and attributes it to negligence on the part of operator or a mischief is done by some body.

In-charge production refutes these allegations saying the previous maintenance was not done properly. This type of argument may be settled, if proper investigation is done to find the cause of repetition of a particular defect in unexpectedly short period of time. Such type of repetition of defect may be attributed to the carelessness of operator.

In practice, it is observed that breakdown or malfunctioning of machine is rarely due to mischief by somebody. Some how, causing intentional harm to machines is considered a bad thing among workman and workwoman. Therefore, it rarely happens.

Job of maintenance workman should be well defined so that maintenance fitters, electrician or a machinist should know very well as to how he should perform. A typical job description is given.

Job Description of Maintenance Fitters

A senior maintenance fitter should perform assigned work in following manner.

- Reach at work place at stipulated time. It may be a fixed time or as told by in-charge. Sometimes, it becomes necessary to reach work place early.
- Take work assignment slip. Understand which machine and at what location is to be attended.
- Go to machine, enquire symptoms of development of defect from operator or supervisor of that section.
- If possible try out to operate the machine and observe or notice the defect yourself.
- Once nature and extent of work is assessed, come back to maintenance section, take necessary tools.

- If necessary, ask supervisor to allot a helper to carry out maintenance work
- Ensure that necessary spare parts are available. Sometimes it is not possible to name spare which are needed unless defective parts or machine element is dismantled.
- Rectify the defect in best possible manner in minimum possible time.
- After rectification try out the machine. Ask operator to produce for trial.
- After operator is satisfied with the performance, inform supervisor production shop and take his OK on repair requisition slip.

Sometimes production machines operator, monitor or supervisor harass maintenance man by pressurizing him to work fast or in a particular manner which is not correct. Sometimes they cut jokes about his work. This all is highly objectionable and should be strictly discouraged. Many a times, maintenance worker is in difficulties as repair or maintenance action does not produce desired result. In that situation, maintenance fitter or repairer is thinking about various ways to solve the problem and sincerely making efforts to remove defect and make machine production worthy.

At this stage, somebody distract his mind then it would be harmful in the interest of organization. Infact there are moments, when supervisor or in-charge of maintenance or production lend a helping hand which boost the morale of maintenance fitters and ultimately completes maintenance successfully.

It is useful to have a performance assessment system for workman of any section of production shop. Assessment may be done yearly which is time to decide increment in salaries to be awarded. A typical assessment form is shown below.

Performance Assessment Form

Assessment period: From .... to ....

Name: Token no:

- *Regularity*: Does he come for work regularly?
- *Punctuality*: Is he punctual in reporting for work?
- *Job knowledge*: Extent of job knowledge
- Ability to work with happy disposition
- Mechanical or electrical aptitude

Contd...

Contd...

- Giving of respect to seniors and co-workers
- Ability to work for long hours if need, without sign of irritation
- Honesty and truthfulness
- Ability of understanding simple mechanical drawings and diagrams
- How clearly is he able to explain or report details of work done?
- Is he ambitious to rise in his profession?

Total marks: ______________

Note for Assessee

- Each question carry 10 marks
- Indicate total marks obtained out of hundred and hence percentage

It would be a good practice to discuss assessment with employee. Sometimes an employee may disagree with certain points of assessments. In such a case, assessor may explain to employee as to where are the possibilities of improving his performance.

A suitable incentive scheme may be linked with performance. In case of maintenance workman, total machinery breakdown time may be one of the measure. A group incentive may be more, if total breakdown time is less. Another additional measure may be as to of what cost spare are used. Incentives may be paid on monthly basis together with payment of salary.

Individual incentives give rise to jealousy and non-cooperation. Group incentives give rise to coordinated work efforts for achieving good results.

In practice, it is seen that one or two workmen try to take advantage of group incentive. They do less work but enjoy full incentive of better performance.

But this type of attitude does not last long. Ultimately supervisor and in-charge of maintenance department form an opinion and hence this type of workman become unwanted.

TEST

1. How many types of machine and plant maintenance are there? Briefly describe each.

2. Name machine elements which might be needed for maintenance and replaçement? Briefly describe.
3. Explain what are fits and tolerances in relation to machinery assembly.
4. Write down hole and shaft tolerance to achieve loose fit, close clearance and transition fit.
5. What do you understand by preventive maintenance of a PVC sheet production plant? What maintenance work is generally done?
6. With the help of two examples describe breakdown maintenance.
7. List out main units of an injection molding machine and sections extrusion plant.
8. Word *Fluid* stands for what? Explain how does fluids are utilized in operation of machines.
9. What are essential components in a hydraulic system?
10. With the help of simple sketch, describe construction and function of a vane type hydraulic pump.

ASSIGNMENT

Note: Tick mark in box for appropriate option. Cross remaining boxes

1. In an enclosed container, hydraulic pressure is
   (a) More on base and less on vertical wall .... □
   (b) Pressure is less on curved surface .... □
   (c) Value of hydraulic pressure is same at all places of surface .... □
2. Which of the following liquids is used in hydraulic system of plastics processing machines?
   (a) Linseed oil .... □
   (b) Lubricant oil .... □
   (c) RO water .... □
   (d) Special hydraulic oil .... □
3. Presence of air in hydraulic circuit causes
   (a) Knocking sound .... □
   (b) Vibration in pipes .... □
   (c) Malfunctioning of directional control valves .... □
4. Hole and shaft tolerances are shown below for location fit. Which one is correct?

(a) Shaft-h6, hole-H7 .... ☐
(b) Shaft-g6, hole-H7 .... ☐
(c) Shaft-c11, hole-H11 .... ☐

5. Screw of an extruder breaks at the shank. It is to be repaired. Which of the following is most appropriate method of repair?
   (a) By using araldite or quickfix .... ☐
   (b) By putting broken pieces together and braze .... ☐
   (c) By putting broken pieces together and weld .... ☐
   (d) After machining alignment holes and putting an insert rod and then to weld .... ☐
6. Which of the following equipments is the part of hydraulic system?
   (a) Change gear mechanism .... ☐
   (b) Chain sprocket .... ☐
   (c) Directional control valve .... ☐
   (d) Relief valve .... ☐
7. Which of the following values of hydraulic pressure is more near to generally developed by a good condition vane type hydraulic pump?
   (a) 1500 psig .... ☐
   (b) 900 psig .... ☐
   (c) 2000 psig .... ☐
   (d) 2700 psig .... ☐
8. Purpose of filters in a hydraulic injection molding machine is
   (a) To divert flow of oil .... ☐
   (b) To reduce oil pressure .... ☐
   (c) To arrest fine solid particles in hydraulic oil .... ☐
9. Purpose of strainer in hydraulic injection molding machine is to
   (a) Arrest course particles .... ☐
   (b) Increase oil flow .... ☐
   (c) Reduce oil pressure .... ☐
10. Which of the following is electrical control in a sheet extrusion plant?
    (a) Integrated circuits (IC) .... ☐
    (b) Over load protector .... ☐
    (c) Timers .... ☐
    (d) Capacitor gang .... ☐
    (e) Solenoids .... ☐

5 Quality Management and SQC Techniques

INTRODUCTION TO QUALITY

Present scenario of industrial products marketing in India and elsewhere is highly competitive. This is due to the fact that large section of consumers have become very quality conscious. It is still on rise due to presence of superior quality products from indigenous and foreign industrial houses. Any industry, medium or small, has to strive for following, if it has to stay with firm footing in the market.

- Competitive pricing
- Excellent quality
- Consumer-friendly service

This chapter is devoted to quality assurance in production processes. Basic principles of quality assurance are dealt here with irrespective of any particular process. In fact, basic principles are generally applicable to most of production processes such as plastics conversion processes.

First of all it is necessary to define quality requirement of a product in terms of its design, material, dimension, tolerances, color, reliability, service life, esthetics, etc.

Once quality requirements are defined, it is necessary to examine various plants and machinery because they have their own capabilities. Some plants and machinery perform various production operation with greater accuracy, whereas some other similar plants perform production operation with a variation beyond limits of dimension and finish, etc. So, one of the important aspects of quality assurance is accuracy and reliability of plant and machinery. Many plants and machinery may work nicely in the beginning but after production for few months, start producing defective quality. For example, a machine having a journal bearing would develop excessive running clearance as compared to ball or roller bearing in place of journal bearing. Another example may be of a steel die which wear out fast, thus producing defective components. A die made of tungsten carbide may last longer before start producing defective components.

Qualities of material or component which are fed to machine are of paramount importance in ultimately producing excellent quality. An example of input component is described here with the help of Fig. 5.1 to highlight importance of good quality of inputs to achieve good quality of output.

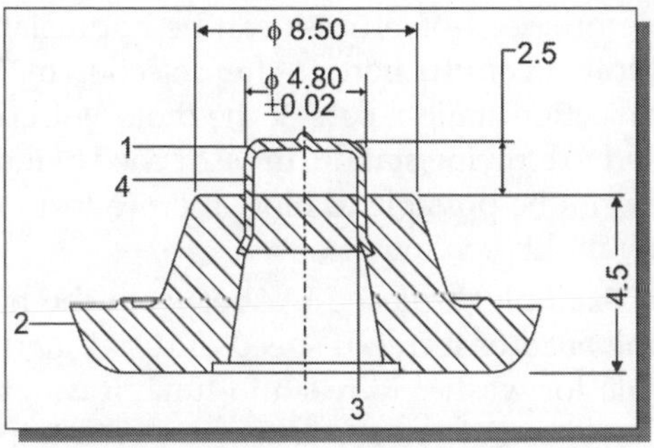

Fig. 5.1: Insert molded component

Components shown in Fig. 5.1 are insert molded. (1) is a sheet metal cap. (2) is thermoplastic. (3) and (4) are those points where no flash is allowed. Gap between cap diameter and mold cavity bore should not exceed 0.03 mm. Clearance above this value will lead to development of flash of plastic. Bore in cavity where cap gets in, does not change within reasonable time of usage. It is the diameter and bore of the cap which may be attributed to formation of flash. It may be appreciated that how critical are the dimensions of cap which is an input to process.

For producing components shown in Fig. 5.1, thermoplastic is an input for insert molding of components. Any variation in melt flow index beyond specified limit may cause formation of flashes. Hence it is highlighted that to ensure excellent quality of components shown in Fig. 5.1, quality of inputs should also be good.

Process Parameters

A production process has certain parameters to suit input material or components. Therefore, to achieve excellent quality of components, production process should also be under pre-determined setting with tolerance. Following may be the parameters:

- Speed of process
- Temperature
- Accuracy in sequence of operations
- Time period of various operations
- Various forces and pressures
- Smoothness of motion, no jerks and vibration

Importance of speed of process can be appreciated with the example of blowing of PET bottle. After injection molding of PET pre-form at injection station, core of the mold with pre-form still hot gets shifted to blowing station. In case speed of indexing slows down, it may not be possible to blow the pre-form fully to take shape of blow mold.

In another example, a plastic washer from the hopper slides down over the head of turret. If speed of turret increases, it may not be possible for washer to reach in time. It can, therefore, be concluded that speed of process is also important. In continuous process, such as extrusion of plastics filaments (thin strands), speed of process is very important.

In many processes, maintenance of correct temperature at various stage of operation is very essential. In glass industry, maintaining pre-determined temperature is necessary to give various shape without development of internal strains. In processing of plastic, maintenance of temperature at various heating zones yields desirable quality product.

There are innumerable examples where maintenance of correct temperature is highly desirable, for example, in sugar mills, milk packing plants and many chemical processes.

In production process, there may be a number of operation which are to be carried out in sequence. Sequences may be mechanized or built-in in the machine. There may be some manual actions also somewhere in sequential operations. Deviation from preset and pre-decided sequence of operation may render product completely unacceptable, that is bad or of defective quality.

Example of possibility of deviation in sequence of operation may be assembly of a few components of an electric switch. It is possible that worker put that component in the die which has to be placed next to another component. Consequently, a defective switch is produced.

In fully automatic machine, like plastics injection molding machines, timings of number of operation are adjustable and programmable. After plastics is injected in mold cavity, some period of time is given (say 5 to 10 seconds) for molded component to solidify enough so that it retains its shape during ejection from mold cavity and core. If the timing is set less by mistake or due to any fault in the system, molded component is ejected prematurely hence making it defective due to over shrinkage or even distortion.

In production process of chemical or paint industries, there may be stages where specified period of time is given for completion of chemical reaction between various chemicals or additives. If this time get increased or decreased, incomplete or defective reaction may be there, thus jeopardizing complete continuous process.

Certain chemical reactions do take place under specified pressure in reaction vessels. If the process is carried out under reduced or increase pressure, resultant output may not be up to the mark. Another example is of plastic blow molded article which is blown in a mold by high pressure air. If the blowing pressure get reduced there is possibility that fine details of cavity surface are not formed on the walls of blown article. In case of hot forging of parts, definite amount of force is required to forge part completely to desired shape. Any reduction in force would not complete the shape of compactness of forged part. Same holds good for coining operation as well.

Smoothness in Motion

In many delicate production processes, smoothness in motion of various machine elements is very important. Non-smooth motions give rise to vibration in machine or a particular unit of machine. Dynamic misbalancing of rotating or reciprocating parts also create vibration. On a miniature lamp sealing machine, glass shells with hanging mounted beads are placed on indexing head. If there is vibration in machine then mounted bead would shake.

Consequently, it may get sealed in glass, when it is not in correct position. In such an event, filament may be much out of axis.

SQC TECHNIQUES AND APPLICATION

Quality assurance of product can be established after the study of production processes. Such studies are undertaken under various conditions.

Case Study

This aspect of quality assurance of product can be very well explained with the help of typical case study.

Figure 5.2 shows a threaded part of household gadget. This part is molded of polypropylene thermoplastics. It is molded by four cavity mold. Molding cycle is six shots per minute, hence producing about 11000 pieces per 8 hours shift.

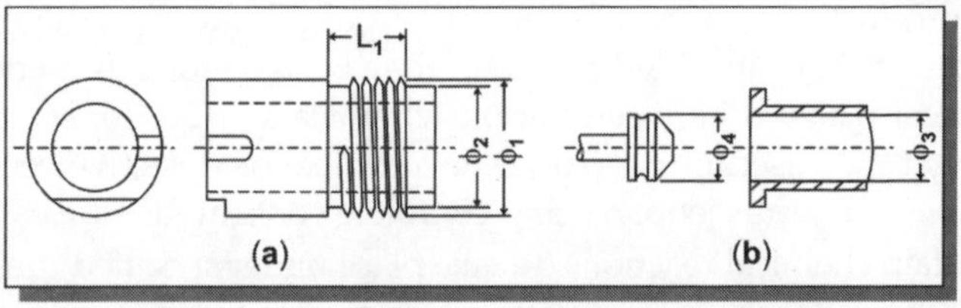

Fig. 5.2: Precision plastics component

Supposing plant is working for 16 hours then production per day would be 22000 components. Component is to be precisely assembled in another component. Designed dimensions are:

$$\phi_1 = 24.50 \text{ mm} \pm 0.05$$
$$\phi_2 = 22 \text{ mm} \pm 0.03$$
$$L_1 = 25 \text{ mm} \pm 0.10$$

Since production speed is high, therefore, it is necessary to know at what time, under what circumstances dimensional deviation crosses specified tolerance limits, rendering a large number of component rejected. Once the pattern of deviation occurrence is known, corrective measures can be taken to eliminate undesirable deviations. First of all samples of component from running production are collected after every hour of production, starting from start of production shift. Let the sample size be ten numbers of components. This means that there would be 80 sample components per shift. All sample components should be measured and mean of reading be recorded as per Table 5.1.

Table 5.1 : Dimensional observations

Date	*Time*	ϕ_1 *24.50 ± 0.05*	ϕ_2 *22.20 ± 0.03*	L_1 *12.50 ± 0.1*
18.08.2007	8.10 AM	24.55	22.23	12.55
	9 AM	24.50	22.20	12.55
	10 AM	24.50	22.17	12.50
	12 Noon	24.50	22.20	12.50
	1 PM	24.40	22.17	12.45
	2 PM	24.45	22.15	12.40
	3 PM	24.40	22.17	12.45
	4 PM	24.50	22.20	12.55
	5 PM	24.55	22.20	12.60
	6 PM	24.50	22.23	12.60
	7 PM	24.40	22.15	12.45
	8 PM	24.40	22.15	12.50
	9 PM	24.50	22.20	12.50
	10 PM	24.50	22.20	12.50
	12 midnight	24.55	22.23	12.55
	1 AM	24.55	22.20	12.0
	2 AM	24.50	22.17	12.45
Min	–	24.40	22.15	12.40
Max	–	24.55	22.23	12.60
Avg.	–	24.47	22.19	12.50

Readings in Table 5.1 are now shown in Fig. 5.3.

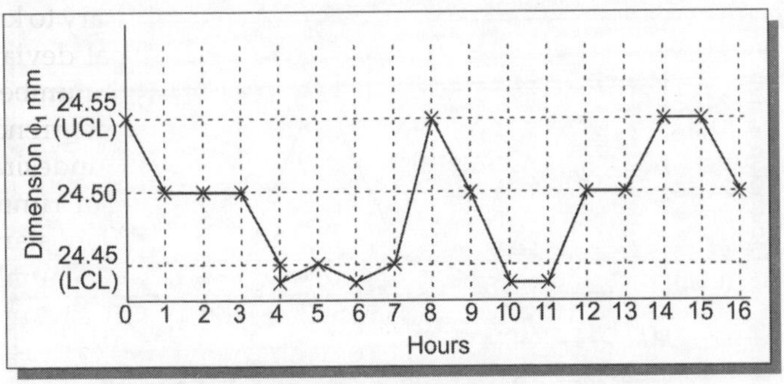

Fig. 5.3: Component dimensions, LCL and UCL

Table 5.2: Dimensional observations

Date	*Real-time*	ϕ_1 *24.5 ± 0.05*	ϕ_2 *22.20 ± 0.03*	L_1 *12.50 ± 0.1*
1.09.2007	8.10 AM	24.60	22.30	12.60
	9 AM	24.60	22.30	12.60
	10AM	24.55	22.25	12.50
	11 AM	24.50	22.23	12.50
	12 mid day	24.40	22.15	12.40
	1 PM	24.40	22.10	12.40
	2 PM	24.45	22.17	12.40
	3 PM	24.45	22.17	12.50
	4 PM	24.50	22.20	12.50
	5 PM	24.55	22.25	12.60
	6 PM	24.55	22.23	12.60
	7 PM	24.50	22.20	12.50
	8 PM	24.40	22.10	12.40
	9 PM	24.40	22.10	12.40
	10 PM	24.35	22.05	12.40
	11 PM	24.35	22.05	12.40
	12 midnight	24.40	22.10	12.50
	1 AM	24.45	22.15	12.50
	2 AM	24.50	22.20	12.50
Min	–	24.35	22.05	12.40
Max	–	24.60	22.30	12.60
Avg.	–	24.47	22.17	12.50

Readings in Table 5.2 are now shown in Fig. 5.4.

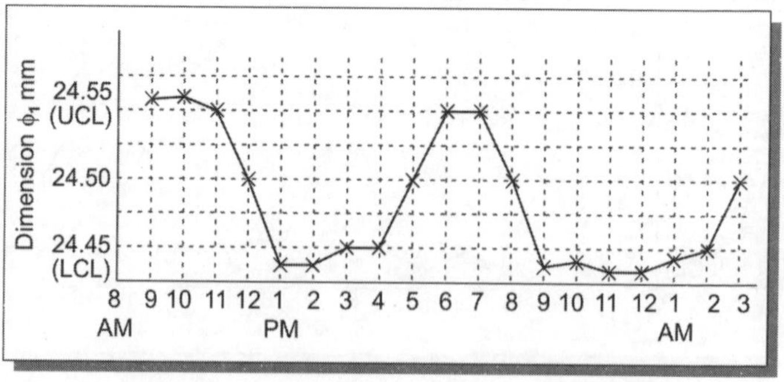

Fig. 5.4: Component dimensions, LCL and UCL

Table 5.3: Dimensional observation

Date	*Real-time*	ϕ_1 *24.50 ± 0.05*	ϕ_2 *22.20 ± 0.03*	L_1 *12.50 ± 0.10*
12.12.2007	8.10 AM	24.40	22.10	12.35
	9 "	24.35	22.05	12.35
	10 "	24.35	22.05	12.30
	11 "	24.40	22.10	12.35
	12 mid-day	24.45	22.15	12.40
	1 PM	24.40	22.10	12.40
	2 "	24.35	22.05	12.35
	3 "	24.35	22.05	12.30
	4 "	24.35	22.05	12.30
	5 "	24.40	22.10	12.40
	6 "	24.35	22.05	12.40
	7 "	24.40	22.10	12.40
	8 "	24.45	22.15	12.40
	9 "	24.45	22.15	12.35
	10 "	24.40	22.10	12.35
	11 "	24.35	22.05	12.30
	12 midnight	24.35	22.05	12.30
	1 AM	24.35	22.05	12.35
	2 "	24.35	22.05	12.40
Min		24.35	22.05	12.30
Max		24.45	22.15	12.40
Avg.		24.40	22.10	12.45

Readings in Table 5.3 are now shown in Fig. 5.5.

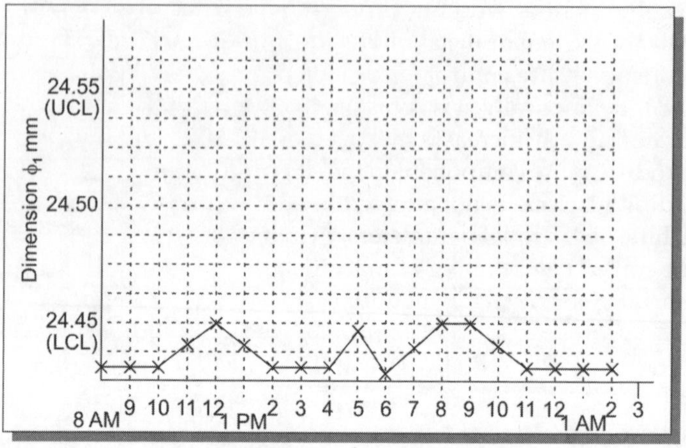

Fig. 5.5: Component dimensions, LCL and UCL

Data in three tables are shown as an example. In practice, sampling is done over a long period of time say, days, weeks, months or even year. Bulk of data helps in analyzing the reasons of deviations. Data may reveal that deviations are erratic or having a definite pattern.

Analysis of Data

Referring to Table 5.1 and Fig. 5.3, it can be seen that dimension ϕ_1 of component has crossed LCL (lower control limit) on two occasions. First crossing of LCL is from mid-day to afternoon and second crossing of LCL is about 7 to 8 PM. Probable reason for such a deviation may be that during day, temperature of mold and injected plastic rises by a few degree centigrades, enough to increase shrinkage of component after ejection from mold.

Crossing of LCL around 7 to 8 PM cannot be due to rise in mold and melt temperature as the ambient temperatures in evening are cool compared to daytime. So there must be some other reason for deviation. It may be increase in supply voltage hence temperature of plasticizing cylinder increases. It may also be due to over enthusiastic operator *playing* with temperature or mold stay setting.

Not losing sight on quality assurance, reasons of deviations should be examined. It may be checked, if mold temperature control system is working properly and with reliability. If not, repair the system or install more effective and accurate mold and melt temperature control. Now look at other deviation around 7 and 8 PM. Practically, it is occasionally found that over enthusiastic operator slightly increases the speed of machine with the intention to produce more component in the shift. If the operator of injection molding machine is not allowed to disturb the setting of various machine operational parameters, then it is indiscipline on the part of operator. Fast running of machine may adversely affect extent of shrinkage of component after ejection.

On examining Table 5.2 and Fig. 5.4, it is evident that some control have become defective at 9 PM and continued till end of shift at 2 AM. On inspecting the machine, it is found that plasticizing cylinder temperature control has become defective. Temperature of front zone is uncontrollably rising. Consequently, plastic melt is at higher temperature than specified and set. Since no attempt is made to compensate for the rise of melt temperature by reducing mold temperature and/or increasing the cycle time, components coming out of mold are on higher temperature than normal. Consequently, dimension ϕ_1 has crossed LCL.

Studies carried out over a period of month or two, can reveal if the defect detected is repetitive in nature or occurred once. If the defect(s) are repetitive then quality assurance policy calls for permanent and reliable repair of system or to replace components or units which are prone to function erratically.

On study of Table 5.3 and Fig. 5.5, it is found that dimension is consistently below lower control unit (LCL). On examining, no fault could be found in the system which may cause consistent defect, that is in all components are below LCL. On further investigation about supplies of polypropylene, it is revealed that the grade is unapproved. It is not the one for which process parameters are already established and put in use. To avoid reoccurrence of such an error, more stringent procedure may be adopted for checking out quality of plastics granules received. First thing is the invoice and information printed on bags. All such actions would lead to quality assurance.

Throw away plastics syringes is another example where quality assurance is of utmost importance. It is not only the dimension important, it should be sterilized properly so that no germs are left clung to syringe. Syringes are then seal packed. Quality assurance starts from factory/works designed for dust free atmosphere. High degree of hygienic conditions are maintained. Syringe sterilization and packing should be fully automatic. At no stage, syringe should be handled by human hands. If at any stage of production it becomes necessary to touch the syringe, it should be done after wearing suitable gloves which are sterilized. So all these arrangements are *built in* quality assurance system. Although testing of syringe for proper sterilization can be tested in laboratory, it may be done by samples from packing.

Lines are periodically tested. Suppose some samples are found to be contaminated then what can be done? It is not like machine which have control for various parameters. May be production has to be stopped. Cleanliness and sterilization process to be examined for possible source of contamination.

Above example is given to emphasize that products of health care, treatment and intervention need *built in* quality assurance environment. Beverage and food products are another example where quality assurance can be attained *built in* production processing system. In dairies, milk after pasteurization should be automatically filled in bottles or pouches without being exposed to atmosphere and human touch to such an extent that there is no possibility of contamination.

CONCEPT OF ACCEPTANCE SAMPLING

Concept of acceptance sampling relates to understanding or agreement between producer and buyer. It is not only be between producer and buyer but between one department of manufacturing concern to another department. Departments may be assembly and dispatch store. Many organizations prefer this system to make doubly sure that quality of product is of acceptable level as per agreement with customer (purchaser). Lots of product are sent to store. Store receives the lot, not accepted as yet. It is accepted only when it is ensured that quality of lot is of acceptable level. Normally, full lot is not checked. Checking is done by drawing certain numbers of product. Quantity or number of samples is decided according to principle of sampling for the purpose of acceptance. Decision of sample quantity normally depends on how reliable production system is? In case whole production system is very reliable then sample size remains small. If system is not reliable then acceptance sample size may be large. In following paragraphs, details of analyzing product are given. After reading, it would provide opportunity to have better understanding of *acceptance sampling*.

Use of statistical sampling is generally adopted when lot sizes are large. As an example, suppose consignment of bottle caps is received in five lots of one thousand each. It was agreed by supplier that one percent of total numbers of caps would be the sample size. This means that out of 5000 numbers of caps, 50 numbers may be drawn. Now, these 50 numbers of caps would be equally drawn from received five lots. It is also agreed upon that out of samples, if more than one defective is found then full consignment may be rejected and supplier would have to take the consignment back for doing the needful.

There are a number of ASTM standards related to sampling. Wikipedia, the free encyclopedia provides sampling related ASTM numbers.

ISO 9000, ISO 17025 AND ISO 14000 STANDARDS

Understanding of ISO standards is quite helpful in formulating quality policy for production of products for customers in country and specially customers from other countries. As already mentioned, International Organization for Standards (ISO) may be considered a source of co-ordination between standards adopted by various countries.

ISO is a big organization dealing in formulating and put into use standards for quality and specifications for products and services. ISO also certifies companies and service organizations that they are operating as per ISO standards. Companies having ISO certification enjoy good reputation. Consequently, other parties and organizations feel safe in dealing with ISO certified organizations.

There may be thousands of ISO standards. ISO 9000, ISO 17025 and ISO 14000 standards are some of them.

ISO 9000: The ISO 9000 family of standards represents an international consensus on good quality management practices. (Source : Wikipedia)

ISO 17025: General requirements for the competence of testing and calibration laboratories. (Source: Wikipedia)

ISO 14000: International standard ISO 14000 from quality network.

www.quality.co.uk/ISO14000.htm

This page provides an introduction to ISO 14000 and ISO 14001, provides links to our many other environmentally oriented pages and identifies official sources of standards themselves. (Source: Wikipedia, the free encyclopedia.)

Management of Quality Assurance (Fig. 5.6)

Any production system is prone to wear and tear, contamination. This may lead to deviation from quality norms. It is highly undesirable. Therefore, suitable actions and precautions are taken continuously on day-to-day basis. It may be termed as management of quality assurance which may consist of following:

- Preparation of quality manual
- Training of workers
- Display of instruction
- Collection of data
- Processing the data on computer having quality analysis software
- To take corrective measure as spelt out by computer and considered appropriate
- To manage that the whole of the work related to quality assurance is done properly and regularly.

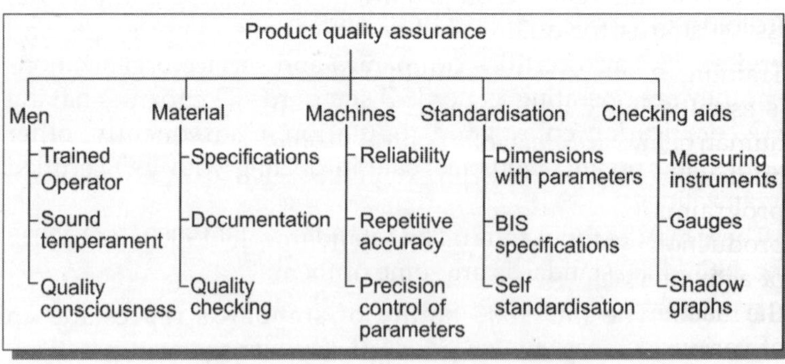

Fig. 5.6: Product quality assurance

Preparation of Quality Manual

Production process may have a number of machinery or group of machines at different places in the plant. For example, fabrication, polishing, plating, vacuum metalizing, assembly and packing. Quality manual may have chapter for each group or section of factory/production shop. Each chapter should contain drawing of component or assembly, description of its function individually or in combination with other components, dimensions with tolerance, input material specification, finish, etc. Manual may also contain instruction regarding testing of component for strength, finish, internal stresses, light fastness, heat tolerance, effect of moisture, etc. whatever is applicable to a particular component.

Quality manual may include sample table, graph which should be filled by quality inspector on specified time or for each lot of specified numbers or quantities. Sample size may also be mentioned. Normally, quality inspectors are not allowed to carry out any adjustment or corrective action. Quality inspector should inform production supervisor by showing him charts or graphs which are indicating quality deviations, in turn, production supervisor may or may not take any action immediately, there has to be a close co-ordination between production supervisor and inspector.

This is very important to clarify here that production supervisor is not expected to put forward the reason for defective quality production as quality assurance examiner did not inform him. It is the responsibility of production supervisor to be vigilant so that excellent quality norms are maintained.

Training of Operators and Workers

Training of operators of plants and machinery and other workers is essential for the management of quality assurance. After all, human beings are created to control the machine. Machines *control* human being only to such an extent for which those are made and programmed by human beings. Plant and machinery of any production process are operated in a definite pattern and sequence of activation. How correctly and efficiently an operator handles the machine or plants depends upon the abilities and knowledge of various controls and sequence of operations. Operator should be capable of understanding various sounds, motions, smells, heat, etc. normally produced in production process. Operator should be capable of noticing if anything abnormal happens. He should also know what quality aspects are acceptable and what are the deficiencies which may render components or assemblies rejected.

If worker or operator is placed to work or control a machine with quick and brief instructions, he may make mistakes, causing loss, damage to plant and machinery and even quality degradation in output from machines. Therefore, great emphasis may be laid on training of workers or operators before putting them on jobs. Training should be in a well-planned manner. Period of training and syllabus depend upon the complexities of production processes.

Training program for an assembly marker or packing worker may be simple and of short duration as compared to training program for machine or plant operator.

Two typical training programs for an assembler or packer are written below.

1. Training programmes

Theory :	15% time	3.5 hours
Practical:	85 % time	20 hours
		Total 24 hours

Theory

- Material and design of packing material
- Differentiation between acceptable and unacceptable materials
- Quality aspects in packing
- Standard output of final packing.

Practical

- Receiving packing material from store
- Inspecting packing material for quality
- Arranging packing material on table or hopper of machine or assembly line
- In presence of training supervisor, packer should carry out packing operation. At the same time listening to training supervisor, he may give guidance for carrying out packing operation
- To carry out packing operation without presence of trainer
- Trainer may call in quality inspector to check packing done by packer
- Packer should be attentive and listen carefully to quality inspector. He may point out some quality deficiencies in packing.

After training of 48 hours, operator may be put to do regular packing work.

2. Training program for a machine operator

Theory : ———— 20% time ———— 9.6 hours
Practical: ———— 80% time ———— 38.4 hours
Total 48 hours

Theory

- Purpose of machine
- Brief working principle of machine
- Construction of machine
- What to check before starting machine or plant
- Safety precautions
- How to start machine?
- How to carry on production?
- Quality requirements of component or product
- To get familiarized with sounds and vibration of machine and plant during proper functioning
- To get familiarized with smells so that if some unusual smell develops, operator should be able to notice

Practical

Practical training may be imparted in the presence of trainer. Operator may be asked to repeat start and stop of machine/plant many times. Trainer shall watch. He should correct the operator, if he is going to make mistakes. In this way, operator would attain confidence.

Operator may now be left alone to run production. However, for few days, production manager or supervisor should keep an eye on quality of product for sometime till good amount of confidence is attained that operator would handle machine and/or plant properly and would produce good quality product/component, taking care of safety precautions.

Display of Instructions

It is useful shop floor practice to prepare pictorial display in language which is generally understood by recruited operators and workers. Display may be a small board fixed near each machine, unit or production line. This helps operator to correct himself. Management may have an advantage that operator or worker cannot find excuses in case something goes wrong with quality norms due to operator not handling the machines as per displayed guidelines.

Collection of Data

Operator may be instructed to keep on collecting components so that about 10 components are collected each hours. Quality inspector may keep on collecting sample components from machine and plant of all sections. Collected samples would be measured and dimensions recorded in attribute nomenclature like poor, good, excellent for finish, for color, shade correct or incorrect.

Collection of data may not be a permanent work. Once corrective measures are taken over a period of time and quality norms are stabilized, quality assurance actions may be redefined. Once this situation is achieved, frequency of collection may be altered according to need.

Processing the Data on Computer

Plant manager, who is managing complete production process, might not be interested in going through all the data recorded by quality inspector. Most likely plant manager would like to know the following:

- Which particular machine is giving much variation in quality?
- Is maximum deviation in quality norms taking place, when a particular operator is handling the machine/plant.
- What is the minimum, maximum and average deviations?
- Most probable reasons for deviation in quality
- Suggested corrective measures.

Since loaded software in computer processes entered data, almost instantaneously plant manager or general manager can see the quality trends of ongoing production on his computer screen his without leaving table.

Corrective Measures

On regularly having a look on quality trends and knowing the probable reasons, a general manager (GM) may order critical observation of a particular unit of a machine, which is most likely to be giving repeated trouble, hence adversely affecting quality trends. GM may decide to replace the unit with a most advanced unit which has proven reliable performance, experienced by other users of unit. Since this type of corrective measure involves expenditure of money, GM may have the authority to go for upgraded unit, if possible to incorporate it in old machine. This does not mean that production manager cannot contribute to the improvement. He may suggest and recommend for replacing existing unit.

In case, on investigation it is found that raw material like plastics granules, aluminum sheets, brass sheets or zinc calots are not up to the mark in quality, do not have qualified quality then corrective measure would be to obtain replacement from supplier. If supplies of raw materials or input components are often not up to the mark then it might be necessary to select alternative supplier.

There is possibility that variation in quality of components or products is due to human error on the part of particular operator or worker. In spite of all automatic controls, there are occasions when operator has to make some adjustments or to stop machine, if output is not of qualified quality.

Components or products having qualified quality, from producer's point of view, may give some performance trouble while in use at customer's place. Simple example may be of precision bolts with hexagonal heads used in assembly of a machine or equipment parts. Sometimes location of bolt in machine is such that for screwing in the bolt takes long time as spanner, open end or ring, can rotate bolt by few degrees only. This may cause irritation to the fitter. In due course of time, this difficulty is brought to the notice of bolt producer. Although bolts are of standard specified quality, then what corrective measure can producer take? He may introduce some suitable quantities of bolts with a slot for screw driver. It may be called a *corrective measure*. It will be possible for fitters to use a screwdriver to screw in the bolt quickly in machine housing and then finally to tighten the bolt with suitable spanner. This means that an additional operation of slot making on the head of bolt would have to be introduced. It means procurement and installation of slotting machine. Is the investment justified or not?, can be decided by a top authority of organization who is looking after marketing, finance and complete operations of the particular bolt producing company.

Regularity in Quality Assurance Related Work

This is an administrative aspect. In a factory or works, there may be a number of sections of production or assembly operations. There may be a few inspector of quality checking to take care of all products. Sometimes one or two persons do not turn up for work. In such an event, production manager might make alternative arrangement. He may place some other persons from office work (say, a data processor) to do the job of quality inspector till absent person joins duty.

TEST

1. What are the factors which keep a plastics product producing company on firm footing in the market?
2. What are quality requirements for a plastics product?
3. How does standard of quality of a product depend on production operations?
4. What quality deficiency may crop up in insert injection molding? Give example with brief explanation.
5. List out process parameters on which extent of quality of a product depends.
6. On what factors, dimensions of extruded sheet and sections depend?
7. On what factors, dimensional tolerance of an injection molded component depends?
8. On what factors, dimensional accuracy of thread portion of blown bottle depends?
9. Dimensional accuracy of injection molded component may vary with day and night time of production. What may be the possibilities of tolerance deviation? Briefly write down.
10. Periodic dimensional checking of components under production may be done and recorded in a tabulated form. Why is it prefer to show recorded values in the form of graphs?
11. Categories product quality assurance and briefly explain each.
12. Why training of operators and quality inspectors is desirable regarding product quality and assuring qualified quality standards?

13. During training of quality assurance department personnel, theory of various aspects is also explained. What are these theory lessons? Name them and briefly explain.
14. In what manner computer can be used in quality assurance work?
15. What do you understand by regularity in quality assurance work?

ASSIGNMENT

Note: Tick mark in the box for correct option. Cross remaining

1. Most likely reason of appearance of *flash* in insert molding is
   (a) High value of MFI .... ☐
   (b) High injection speed .... ☐
   (c) Too much clearance between cavity and insert .... ☐
2. Achievement of qualified quality of vacuum formed product depends on which of the following?
   (a) Density of plastics sheet .... ☐
   (b) Speed of forming .... ☐
   (c) Thickness of sheet .... ☐
   (d) Light transmittance .... ☐
3. Quality assurance depends on
   (a) Human factor at all stages of production .... ☐
   (b) Quantities to be produced .... ☐
   (c) Plant and machinery reliability .... ☐
4. Which of the following statements is correct?
   (a) Quality attributes are difficult to specify .... ☐
   (b) Measurable quality factors are difficult to express .... ☐
   (c) It is possible to check quality of component or product by gauge and measuring instrument .... ☐
5. Effective quality management includes which of the following decisions regarding quality to be left on?
   (a) Production in-charge .... ☐
   (b) Induction of quality checking and recording team .... ☐
   (c) Quality parameters to be well defined .... ☐
6. Use of which of the following would ensure production of product with qualified specifications

(a) Drawing of product ☐
(b) Photograph of product ☐
(c) Use of gauges ☐

7. Which of the following are normally used on shop floor for quality checking?
   (a) Model of product ☐
   (b) Duly marked sample component ☐
   (c) Microscope ☐
   (d) Gauges ☐
8. Which of the following is necessary for *in built* quality assurance?
   (a) Display of quality norms near machines ☐
   (b) Neat and clean environment ☐
   (c) Preventive maintenance of plant and machinery ☐
9. Which of the following are quality control aids?
   (a) Testing equipments ☐
   (b) Shadowgraphs ☐
   (c) Measuring instruments ☐
10. Which of the following time distribution for training would you prefer in the ultimate interest of maintaining good quality of product?

    Percentage of time distribution

Theory	Practical	
15	85	☐
50	50	☐
20	80	☐

11. Use of computer in management of quality is
    (a) Not necessary ☐
    (b) Preferred ☐
    (c) Essential ☐

6 Costing and Industrial Management

BASIC PRINCIPLES OF COSTING

In this part of chapter, basic principle of product costing is dealt with for policy makers in top management. For a small scale industry, top policy maker may be a proprietor or partner. Large industries which are generally limited companies, policy making body is board of directors. Board discuses and decides as to what is to be the target of profits for next 3 to 5 years. Once it is finalized and passed by board, then it is passed on to managing director to formulate line of action to achieve target set in policy.

There may be a number of action plans, such as:

- Reduction in production cost of existing products
- Value addition to existing products
- Introducing new products
- Austerity measures on expenses
- Diversification plans.

In this part of chapter, only reduction of production cost of existing products is discussed. Management could have possibly decided to increase selling cost structure to increase profit. But such a step is not always possible due to resistance from market for increase in selling price, because such products from other manufacturer are available at a certain price. If price is increased by a particular manufacturer, then sales are expected to go down. Sometimes plastics product manufacturer's association steps in and after many deliberations product manufacturers decide for a common raised ex-godown price for a particular product. But this rarely works due to variable market conditions from place to place. Consequently, it comes to reduction in production cost without increasing ex-godown price for increasing profit. Ultimately, it comes upon works team to analyze costing to find out as to where cost reduction measure may be taken up.

Basic principle of cost reduction is either to increase production quantities per shift or to reduce expenses incurred on existing production level. Both the options are quite difficult to achieve. Increase in production per shift may be done by better supervision so that unauthorized stoppage of machines does not take place. Generally, such stoppages are very little, say 15 to 20 minutes in a shift. Supposing, by any arrangement stoppage of 20 minutes is eliminated then the increase in production would be about 0.4%. Increase would nominally reduce production cost. It is not going to add much to the target of overall cost reduction as set by management. It is not advisable to increase the speed of machine as it is likely to lower quality standard. So the only alternative left is to increase number of cavities, say from 4 to 6. This action would increase production by 50%. Getting a new six-cavity mold would need an investment. Works manager may prepare a comparative costing of present mold and proposed mold, and may send to finance department to examine the proposal for procuring a new 6-cavity mold.

A typical proposal sheet is provided below to study as to what information and in what manner are placed.

Proposal Sheet

From: Works manager To: Finance dept.

Date: 25.Sep.2011

Proposal: To get a six-cavity mold made to replace existing four-cavity mold

Objective: To reduce production cost substantially by increasing productivity

Production per shift by existing 4-cavity mold

- Let specified number of shots per minute 5
- No of components per minute

No of cavities × shots per minutes

4 × 5 20

- No of components per hour

20 × 60 1200

- No of component per 8 hours shift

1200 × 8 9600

- Similarly, production with 6-cavity mold
  $6 \times 5 \times 60 \times 8$ 14400

  Hence the increase in production is
- 14400 – 9600 = ... 4800

  %age increase 50%

  Cost factor (provided by finance department)

  Machine and mold cost per shift Rs.5000/-

  Operator cost per shift Rs.800/-

  Helper cost per shift Rs.500/-

  Overhead per shift Rs.10000/-

  Hence total expenses per shift Rs.16300/-

  Processing cost per component with 4-cavity mold 16300/9600
  = Rs. 1.69

  Processing cost per component with 6-cavity mold Rs.1.13

  Hence reduction in processing cost per component = 1.69 – 1.13
  = Rs.0.56

  Reduction in processing cost per shift .... 0.56×14400
  = Rs. 8064

  Reduction in processing cost per 1000 components
  = 0.56×1000 = ... Rs. 560/-

  Cost of 6-cavity mold .... Rs. 3 60 000

  Hence number of components require to be produce to recover mold cost ....643 thousands or 643000 (approx)

  Production per day on 2 shifts basis .... = 28800

  Days (on 2 shifts) needed to produce = 22 days, say one month

Note: On one shift basis, mold cost recovery will be in 44 days. Say two months.

There would be slight saving in material, may not be taken into account in working out recovery time.

STANDARD AND MARGINAL COST

Logic of costing of plastic components or items is given in Fig. 6.1.

Costing of Injection Molded Components

Costing may be done in following manner:

- Standard
- Marginal

Standard costing is based on taking each factor into account. Value of each factor in terms of money is taken into account which are prevailing at the time of costing. Standard cost has two branches, i.e. material cost and processing cost.

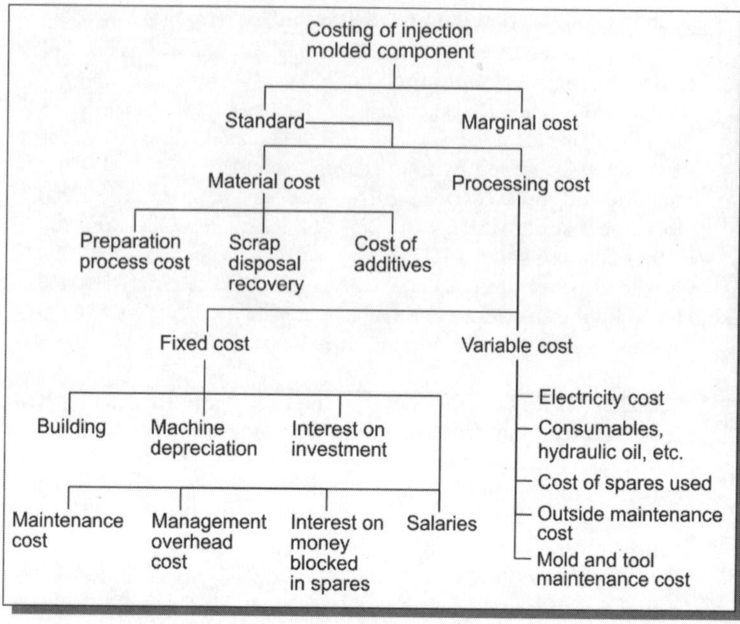

Fig. 6.1: Costing of injection molded components

Let material be plastics resins, granules, powder, as it is called differently. Plastics granules may be directly used without under going any preparation process. Preparation of material means adding master batches, additives, etc. and mixing in tumblers or mixers till all plastics granules are homogenously mixed with additives. This process incurs expenses in term of labor, electricity and machine depreciation cost. For calculating preparation processing cost, per hour machine cost and electric cost for one hour operation must be known and recorded. This information is repeatedly needed whenever costing of a particular product is done. Similarly labor cost per hour should be known as standard and recorded. These data are generally provided by account or finance department of the company. Landed cost of various additives, master batches and modifiers should also be known so that production works is able to do initial costing. Cost of plastics as landed in company store should be known in term of per kg or per bag of 25 kg. Adding all cost would give standard material cost. Most likely there would not be significant quantities of scrap to take its disposable value into account. In case process is yielding substantial quantities, then scrap may be sold and money recovered has to be deducted from calculated material cost. Now the resultant cost may be considered as final material cost for product quantities produced. Product means, injection molded, blow molded, vacuum formed, extruded sheets and films and fiber reinforced items.

Now processing cost may be worked out to be added to material cost so as to get final product cost before any auxiliary operation like marking or packing operation is done. If necessary, costing of auxiliary operation can also be done.

Processing cost mainly consists of fixed and variable costs. Fixed costs are those which are not dependent of variable factors of production. Fixed cost is that which does not change every now and then. In fact values of fixed costs may change once or twice an year, when management carries out some improvement and modifications in existing buildings, etc. Further, investment in advertisement for a period of one year or so may enhance fixed cost. When there is possibility of change then a fixed cost is what?. Answer to this question is that once values of fixed cost factors are enforced by management, they are to be taken into account when doing costing for quite a long period of time, say 6 to 12 months. Under fixed cost, following factors are included.

- Building
- Machine depreciation
- Interest on investment
- Maintenance cost
- Interest on money blocked in shape of spares
- Salaries plus perks
- Management overhead costs

Whole premises of a manufacturing facility consist of total area of land, factory shed, office building, infrastructure, essential service installation like water hydrants. All these may have cost in millions of rupee. Management decides that all these costs under heading of *building* must be returned in say 5 to 8 years. Then the total amount to be got back is divided into year and month. Suppose monthly recovery of this be X. Now this X value has to be further divided among various production shops according to their area. Let us assume that Y is the *building fixed cost* share of injection molding machine shop. Then it can be further taken into account according to production quantities. Ultimately fixed *building cost* may be worked out for one thousand components. One thousand is an arbitrarily figure which is considered a unit of product quantity, say 1000 pieces of a mechanism gear.

Machinery and auxiliary equipments are purchased by organization with investment of money. This investment must be got back in reasonable time to replace machines, which have completed their service *life*. With passing time, prices of machines increase, hence escalation factor is also considered while calculating depreciation. Generally, recovery period for investment is taken as 3 to 6 years. Hence the worked out figure is the depreciation for particular machine and auxiliary equipment attached to it, say robotics. This comes under fixed cost.

Industrial organizations generally invest on machinery or for expansion of building by taking loans from banks. These loans with interest are returned to bank in installments. Return of interest and principal amount is done in stipulated period of time in years.

So for a year or more this return of loan together with interest is a fixed expenses till pay back is completed. Hence it is added to fixed cost.

Maintenance of plant and machinery is carried out by regular and hired labor. Salaries of regular maintenance employees are paid monthly. In addition to this, they may get bonus by the end of financial year. Although bonus may vary in amount every year but management normally keeps a fixed fund every year for disbursement of bonus as deemed fit. Plus or minus balance from fixed budget is then adjusted in the books of account. Therefore, allocation of this fixed amount is taken as fixed cost.

Other fixed expenses for maintenance may be depreciation cost of tools, jigs, equipment and cost of floor area which maintenance department occupies.

In big production factories, a number of critical and non-critical spares are stored. Cost of these spares may be in lacs of rupees or even more, depending upon number and type of machinery. For example, spare set of calendaring rolls, extruder screws, pumps, valves, etc. are very costly items.

Total cost may reach million of rupee or even more. If cost of other items like hose pipes, contactors and other electrical items is added together, then it may touch more than million of rupees hence a big amount of money is blocked in the shape of spare parts. Total value of spares may keep on varying from month to month but for the purpose of accounting an average value is taken as fixed maintenance cost.

For all practical purposes of accounting, salaries, perks, group insurance of employees, allocation of funds towards employees provident fund are almost fixed expenses, at least for a year. Normally, every year some increment in salaries are awarded by management. This makes slight change into fixed cost which may be taken into account while fixed maintenance costs are worked out.

Management overhead costs include a number of expenses towards following:

- Statuary payment to government departments
- Budget allotment to sales and marketing organization of the company
- Towards advertisement
- Contingency funds
- Security system
- Towards safety mechanism for company properties, plants and machinery, employees and visitors
- Vehicles
- Environmental control, gardening and landscape upkeep
- Expenses on tours, conferences, and training programs for technical and office employees
- Office equipment like computers, xerox machine, cabinets, etc.
- Allocations of funds for dividend to share holder
- Funds set aside for research and development
- Legal expenses
- Salaries and perks of directors, senior officers, office staff and others
- Canteen, rest rooms, first aid room and washrooms for factory workers and office employees and for others.

Almost all small, medium and big industrial organizations are proprietary, partnership, private limited or public limited. These organizations are registered with a number of government organizations such as industries department, income tax, sales tax, pollution control board, municipality, fire fighting department, transport department and even police.

Companies pay sales tax and income tax on regular basis. Extent of taxes depends on actual value of total sales and income of the company. It is normally done at the end of financial year or as the case may be. Certain other fees are to be paid to other departments. Depending on the size of companies, industrial houses and their policies, set up of sales and marketing depends. Some companies have their own marketing organization. They have their branch offices and depots at different places in the country. Many manufacturers of *Brand* products have global presence with their branches or representatives located at strategic location in various countries. So all the expense in setting up and maintaining branches, depots and representatives is a fixed cost.

There are companies who do not have big sales or marketing organization. They engage outside marketing agencies who do the marketing for retaining company. This arrangement is considered to be more hassle free and cost less. So expense incurred in retaining agencies is also a fixed cost, at least for a stipulated period of 3 to 5 years. Obviously terms of agreement may keep on revising. So it may be considered an yearly fixed cost.

Nowadays, many companies depend more on aggressive advertisement through electronic media by which they *reach* millions of prospective customers. There may be parallel advertisement from other companies for almost the same product. Now it depends on viewers as to which brand they would prefer. It very much depends as to what impact does an advertisement leaves on viewer. As compared to some other countries, in India *brand loyalty* is quite strong. But with coming up young generation, cost and performance factor is on fore and *brand loyalty* is taking a *back seat*.

Let us analyze example of two ads on TV. Producers of *MDH masala* advertise their range of product very frequently on TV. Although masalas are low priced items but they are spending a lot of money on TV ads. Result of their advertisement must be that large quantities are sold throughout the country.

Take example of another performance-based advertisement of *Sensodyne* toothpaste which graphically show to viewers that it block the sensation of pain. Graphics are so designed and presented that they instantly catch the attention, in spite of *brand loyalty* for other toothpaste. One would like to try out to see, if Sensodyne provides some relief from pain. There are high profile ads for various brands of automobiles. Generally, ads for automobiles must be quite expensive as these ads are video based.

Contingency funds are those which are needed to meet expenses for unforeseen or unexpected circumstances which may come up, for example, a natural disaster or an accident or a plant shut down due to technical reasons and off loading the work till the problem is solved.

Expenses are incurred in engaging outside agency and internal security setup. Both expenses are fixed in nature. Employed security personnel are generally given regular salaries. Residential quarters are normally provided in premises. Uniforms are provided. Some security personnel are provided with fire arms. Flood lights are installed at strategic position. All these paraphernalia are also a source of fixed expenses.

Company may have a number of properties in one or a number of cities, depending upon nature and size of company. In any case, company has to put a mechanism in place to take care of security aspects of properties. Consequently, incurring fixed expenses.

Companies may have a number of passenger cars, staff cars and transport vehicles. These vehicles are also source of fixed and variable cost which may be taken into account while doing costing of a product.

Environmental control includes installation of waste disposal plants, general cleaning. It is also a good practice to maintain lawns and garden. This provides freshness in air and pleasing views. This gives a good effect on the feeling of goodness on the employees. This type of environment leaves a good impression on visitors. Expense on environmental control is almost of fixed nature.

To keep the wheel of business rolling smoothly, fast and efficient, it is necessary that officers and staff visit parties and branches as and when required. For this, they have to tour. Senior officers or general managers or managing directors have even to go abroad in connection with negotiating and finalizing a business deal. Tour may also be for meeting with the heads of other company for collaboration.

Sometimes technical personnel have to go out in country or abroad to sort out some technical problems or to gather information and knowledge about new technologies. It is generally attained by visiting international fairs like *Kunstaff messes* a plastics fair in Germany. It is very famous. In fact many a times visit to such fair gives dividend by way of implementing new method in existing process techniques. To give an example, many years back author had designed a 12-cavity mold for a small technical component with conventional direct gate. It was necessary to de-gate component manually. A visit to *Kunstaff messe*, ideas of submarine gating was picked up and a new 16-cavity mold with submarine gate was made and put in production. This eliminated manual de-gating consequently clean, fine automatic de-gating could be achieved on regular basis.

So such visits also incur expenses. Management normally keep a fixed budget for this purpose.

Conferences and technical training programs are sometimes necessary to keep the performances of employees up to date and competitive.

Another source of fixed cost is office equipment like computer, xerox machine, cabinets, tables, chairs, etc.

Depending on financial performances of company, specific fund is kept aside for giving dividend to shareholders and bonus to employees. This is also a fixed cost for a particular years.

Many big industrial houses have their own research and development departments which are assigned with the responsibility of improving the quality of product and cost reduction through alternative plastics and innovations in production process.

Salaries and perks of directors and senior officers are also a fixed expense, say over a period of an year. Next year there may be some changes. Then too, for that year expenses would be fixed. These expenses are also a part of *overheads*.

Many companies maintain canteen, rest rooms, first aid room for factory workmen, office employees and others. It is generally necessary with factory which is located at remote places and two or three working shifts are there. This setup is a regular source of fixed expenses.

Reverting back attention to costing diagram, processing cost also has *variable cost*. Electrical consumption for running machine and equipment depends upon the time period for which production is carried out. Hence consumption of electricity keeps on varying as requirement of product quantity may keeps on varying. Consumables like hydraulic oil, cleaning items may also vary depending on production quantity and condition of machinery. Use of spares is not a fixed thing. Spares may be or may not be used over a period of say one to five months. So this expenditure is also variable and may be taken into account while determining processing cost. Engaging outside agency for maintenance is not a regular job. It may be needs once every 2 to 3 years.

Expenses on mold and extrusion dies repair are not a regular feature. It is done when needed. Fitters, mold makers and die makers may be regular employees then their salaries are a fixed cost and are not considered in variable cost. In this case, variable cost would be use of spare and fastners, if any.

Marginal Cost

Marginal cost in a production is a topic under cost accounting which has a variety of influencing factors. It may need an exhaustive treatment to explain various aspects. At present, it is beyond the scope of this book to deal with this topic in detail.

However, elementary idea about marginal cost is provided.

For understanding definition of marginal cost, it is essential to remember what a break even point in production is. Break even point is the quantity of units production where production cost equals selling price of product. This means that there is no profit or loss. It is because of the fact that fixed cost remains the same. Only variable cost increases directly related to production quantity remains. On crossing break even point, cost of additional one unit is marginal cost.

Let us assume that 1000 cosmetic bottles are a unit quantity. Table 6.1 shows effect of increase in output on marginal cost.

Table 6.1 is also represented by a graph which is presented in Fig. 6.2.

Marginal cost is an increase or decrease in total cost of production affected by producing one additional unit of item.

It can further be illustrated by an example that the total cost of producing one bottle is Rs. 5 and the total cost of producing two bottles is Rs. 9. Then the marginal cost of increasing output by one unit is Rs. 4 (9 – 5 = 4).

Table 6.1: Cost date

Unit of output	*Total cost, Rs*	*Marginal cost, Rs*
1	40	40
2	70	30
3	50	20
4	80	30
5	120	40
6	170	50

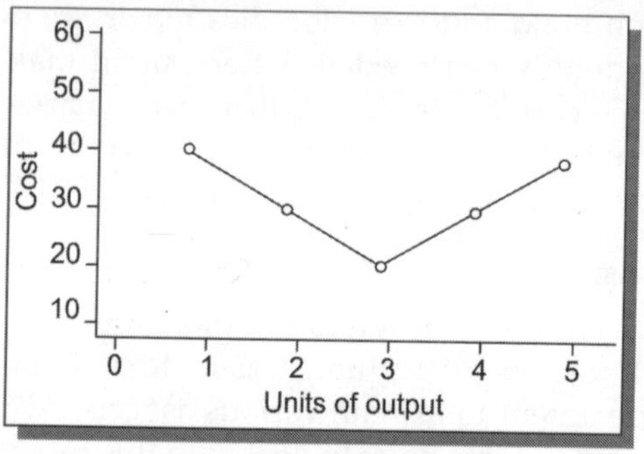

Fig. 6.2: Output versus cost

PROFORMA FOR COSTING

Proforma Costing Sheet

- Name of component:
- Weight in grams:
- Type and grade of plastics:
- Process type:
- Landed rate of plastic:
- Weight of plastic per component after including share of scrap:
- Number of cavities:
- Shots per minutes:
- Total production quantity per shift:
- Machine fixed cost (FC) per shift (*):
- Machine variable cost (VC) per shift (*):
- Material cost per shift (*):
- Total cost (CT) per shift (*):
- Total production quantity (QT):

- Per unit cost CT/QT = :
- Date of costing:
- Costing done by:

Note: Cost factor marked (*) may be obtained from accounts/ finance department of organization.

INTRODUCTION TO INDUSTRIAL MANAGEMENT

An industrial organization has to perform multifarious activities such as decision-making, framing action plan, distributing responsibilities, putting in place feedbacks and control mechanism, planning short- and long-term provision inputs like finance, plant and machinery, material, manpower, to run organization profitably, to setup and manage marketing and sales organization, co-ordinate interaction between various departments and also with outside agencies and govt. organizations, ensure that statutory obligations are fulfilled regularly, strive for good dividend to shareholders, welfare of employees, fulfillment of social obligations towards society at large, keep the foundation of organization intact and strong, manage the organization in such a way that it enjoys an excellent reputation in national and international markets.

Small industrial organizations are normally managed by only proprietor or partner and a few employees. In practice, it does not need a formal organizational structure. In case of big industrial houses with multifarious activities, a formal organization structure is needed to manage. A typical organization structure is shown in Fig. 6.3 for a public limited company having an yearly turnover, say above 100 millions of rupees per year.

Above shown organizational structure seems to be a *top heavy*. But it is necessary as experts of various activities are needed due to intricacies involved by virtue of big organizations. One or two persons are not expected to perform all functions efficiently.

PERSONALITY

Personality of an individual is an important factor while recruiting an employee. Personality very much differs from one person to another. Various jobs need different personalities.

Personality does not specify any one quality of a human being. In fact person has a number of qualities in varying degrees. When we talk of personality in context to industrial employment, then

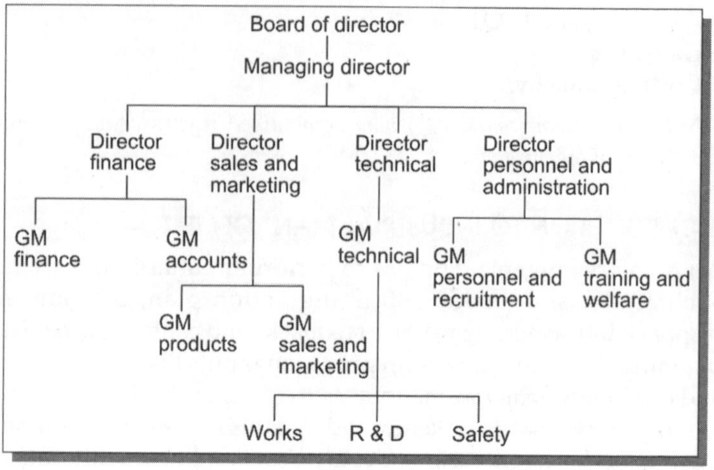

Fig. 6.3. Industrial organization

predominant quality or ability is *personality* which an employer looks at. For example, when a security officer is to be recruited then his physique and looks are also observed. Apart from other things, he should look physically strong and dominating.

While recruiting a works manager, his personality should *speak* about his technical knowledge, experience, quick and firm decision maker, ability to reason out actual facts, humane approach and all important leadership quality.

An employee for the post of research and department scientist needs not be a physically strong person. He should be cool minded, ability to do deep thinking, ability to observe things critically, temperamentally does not give up easily in case things do not turn out as expected.

Recruitment of an employee is itself a critical job. Sometime to judge a candidate for employment for a particular job may need a psychologist to be the member of interview board. For selecting a senior officer or a director, senior directors of board or chairman of the company may interview prospective choice.

APTITUDE

Aptitude of a person is his liking or desire to get involved in a particular activity or work. It is an inherent quality of a person which develops from childhood. Some children not only play with mechanical toys but also seen breaking down the toys. When they grow young, they like to get involved in mobiles, cars, household gadgets. Persons with good mechanical aptitude are generally found to be suitable for supervision or operation of plant and machinery.

There are people who have good aptitude for electrical or electronic system because from their childhood they have played with switches, wiring, dry cells for lighting miniature lamps or making electricity with a bicycle dynamo, etc. There are people who have played with radio transistors, tape recorders, etc. and attained a liking and inclination to get involved in electrical and electronic system. On account of basic and specialized education and training, they become good technicians, engineers, etc.

It is observed that while children are playing, one of them dominates on others and guide other fellow children to play in a particular manner. This sort of behavior persists when they reach schooling stage. There, they take part in extra curricular activities like participation in drama, sports and organizing a function. When they reach college level, their activities become of a higher level such as participation in students union's activities. All these indicate that the person has an aptitude of leadership. This aptitude of an employee may be an asset for employer. He may be an effective incharge of a welfare department which cares for well being of workman and workwomen and see that they work with happy disposition.

MOTIVATION

Motivation is a technique for making an employee or a group of employees to perform their work in more effective manner. Getting a certain performance from employees by order and giving punishment may not be lasting as they do it out of fear and a feeling of hatred. If they are explained about situation and talked around to make them agreeable to work in difficult situation. To do this, many a times in-charge of group of those workmen have to work with them shoulder to shoulder. This attitude of an incharge *sends* a message that *I am here to help you in completing or performing a job in difficult situation*. Author quotes a situation when there was a short fall in production quantity due to late arrival of plastics. Normally, machine is stopped for lunch time. But operator offered to continue working without taking lunch. Author at that time was overall incharge of plastics production. He told operator to go for lunch as he would attend to machine, and he did it. Such type of motivational technique leaves a long-lasting effect.

COMMUNICATION

In industrial environment, importance of communication cannot be undermined. Most of the communications are related to work. Any mistake or misunderstanding may cause loss of time and even damage to machine or system.

There are four ways in which communication between two individual or group may be done, which are as below.

1. Verbal
2. Written
3. Sign language
4. Drawings

In day-to-day working, most of the time, communication between workers, supervisor, incharge or others is verbal. One has to be careful in communication so that correct instructions are conveyed. Ambiguous communication may lead to wrong action which may even lead to damage to machine. An example is given below.

In-charge of plastics molding section tells an electrician that while doing wiring, make sure that motor of hydraulic pump rotates in clockwise direction. Electrician follows the instruction. Does wiring but there is no indication of development of hydraulic pressure in the system. Electrician come to incharge and inform about the matter. In-charge goes himself and finds that motor is rotating in anticlockwise direction. He says to electrician that he has done wiring in wrong way consequently direction of rotation is anticlockwise. Electrician says that *no sir, see motor is rotating in clockwise direction.*

Why this difference in perception? It is because that each of them is viewing the motor from different direction and this happened due to incomplete communication. If in-charge would have *told electrician direction of rotation of motor should be clockwise when seeing from pump side.* With this communication, there had been no mistake.

Written communications are generally confined to, e.g. giving an application for leave, giving a written weekly schedule to monitor, operator or technician. It may be passing a maintenance demand slip received by in-charge maintenance from in-charge production shop. Sometimes a new employee is given a guidance write up which contain Do's and Don'ts.

Sign language for communication is used when there is too much of sound in workshop and it is difficult to hear from a distance. Suppose some body wants to call a colleague to him then he gives a usual signal by hand to come. Another example may be a sign language communication between a overhead traveling crane operator and a colleague on ground directing crane operator about movement. Suppose man on ground wants that crane is stopped to travel then he raises his both hands and then turn thumb of his right hand down, indicating that the load which crane is carrying should be lowered. There may be many examples of sign language communication.

Communication by drawing is generally resorted to where some parts of machine or tool are to be made new or to be repaired in tool room. In this communication, machining symbols, dimensions, etc. are indicated. It is our common experience that while explaining a address to some body, we sometime draw a map on a piece of paper and give it to the person to whom address is being explained.

LEADERSHIP

Leadership is a complex quality in an individual. This quality in a shop floor supervisor and a director of a company would be quite different. A supervisor has to establish his leadership by his technical and supervising abilities. He may have to shout sometimes, move briskly around workshop. Say yes or no in a decisive manner.

Leadership in a director need different set of qualities such as being well-informed about industrial trends, company's performance, ability to analyze facts and figure and draw correct conclusions, ability to convince others for an appropriate decision making, to handle corporate activities in a profitable manner. All these and other qualities added together generate quality of leadership.

ORGANIZATIONAL STRUCTURE

Human beings have learnt to organize themselves from ages, when they had to fight for their existence and to save themselves from huge and wild jungle beasts. By experience, they learnt that united efforts are much more useful. As a result of this realization and development of civilization, united efforts gradually took shape of formal organization in management of civil affairs, factory production, military and allied services like travel, post, health care, etc.

In today's modern society, functions of various organizations have become very complex due to multifarious activities and their size. Not only this, unavoidable interactions between many organizations have made development and creation of *Organisational structures* foremost necessity.

Organizational structure defines as to which entity has to be at what place, to which other entity one has to report and who are to report to a particular entity. Authorities and responsibilities are defined for each and every member (entity) of the organization. As a result of this, works are performed in a disciplined manner and deliver result-oriented output.

Organization may be small, medium and big in terms of number of persons involved, number of product, services rendered and turn over in terms of money or other output.

A small company or a shop producing edible mustard oil might be having a few persons and turn over in terms of money may be a few thousands of rupees. On the other hand, a small company polishing and finishing the diamonds may have few employees, but turn over in terms of money may be huge. A big production organization might be producing a variety of high cost products like electrical machines, control equipment, refrigerators, washing machines, deep freezers, water purifiers, etc. Such organization might be having a turnover in million of rupees.

Let us take example of educational institutions where thousands of young boys and girls pass out every year. Generally, there are many branches and sub-branches of educational disciplines such as science, arts, commerce, engineering and technology, medicines, petroleum technology, oceanography and many others.

It really needs a very complex organizational structure to effectively take care of all the activities and ensure that good education is provided. Not only this, an infrastructure has to be managed for the welfare of student, faculties and staff, etc.

Generally following types of organizations are adopted:

- Flat organization
- Hierarchical organization
- Matrix organization/management

Flat organization is sometime known as horizontal organization. It is generally practiced in small organization where a proprietors or partners directly supervise work of few employees or daily wagers. At the same time, they also deal with customers and outside statutory agencies under which organization comes, for example, industries department, income tax, municipal shop act registration, etc. Since activities are not too complex, it can be handled by proprietor. In such type of organization, there is a direct interaction between employers and well-experienced workers. In such cases, workers can also participate while decisions are taken by employers. Employees have a feeling of great responsibility as the outcome of their work is directly seen by employer, on whom progress of employees depends. A typical flat organizational structure is shown in Fig. 6.4.

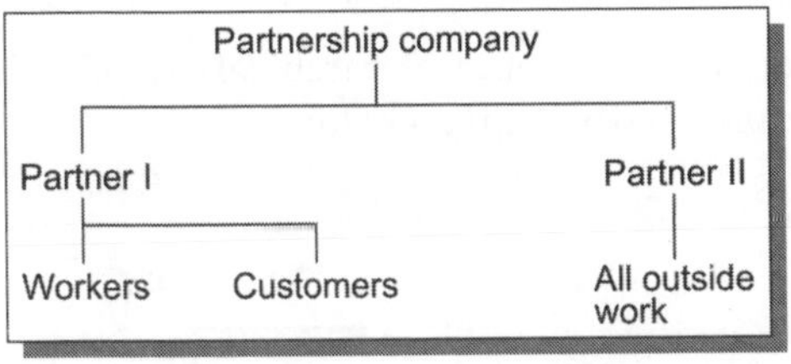

Fig. 6.4: Partnership company

In hierarchical organization, an entity (formal/official position) may report to one or none. One or more than one may be reporting to a particular entity. In the last, a member may be reporting to one but no one reports to them. In this type of organizations, entities have a tendency to control reporting entities. It may be counter productive, if domination tendency exceeds to certain limits.

Hierarchical organization is generally represented by Fig. 6.5.

Apex is the top most position in hierarchical organization. It may be occupied by a managing director whose responsibilities are enormous. By virtue of this top position, enjoys higher status, salaries and perks as compared to other officer reporting to MD. In hierarchical organization, sometimes there is unusual situation, when an employee, say of a manager rank official reports to work manager but at the sometime he reports to product development head for taking technical instruction about developmental activity which takes place in jurisdiction of works manager. In such cases, manager has to be very careful and tactful so as to avoid any working friction between two heads of different departments.

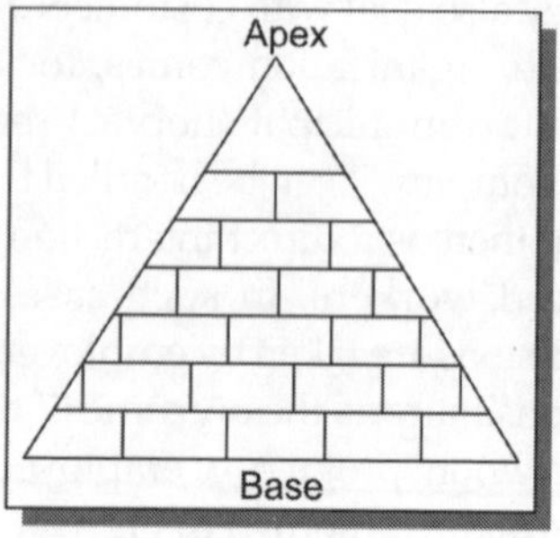

Fig. 6.5: Apex organization display

To elucidate the concept of hierarchical organization, a typical Fig. 6.6 for a university is given below.

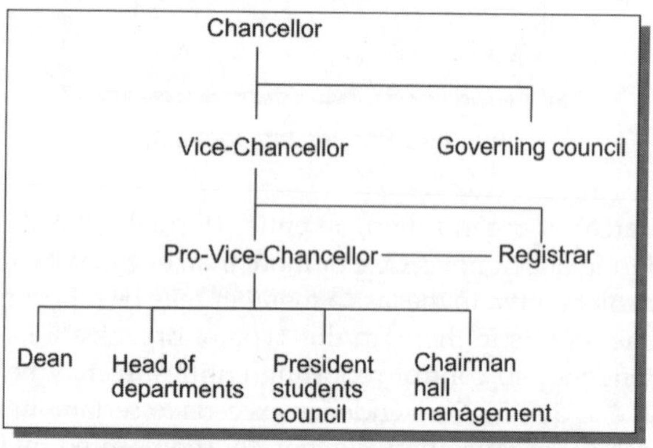

Fig. 6.6: Another organization structure

Another typical example of hierarchical structure of a production organization is shown in Fig. 6.7. This organization produces, say three products A, B and C in bulk quantities. It has its own sales and marketing organization. There is a department which continuously does research and development work to improve utility and quality of existing products and exploring possibilities of introducing new type in present products A, B and C.

There are thousands of employees, including base workmen and workwomen. Outside agencies are also engaged as ancillaries. Company is of limited status with public shares. Keeping all above in mind, a typical hierarchical organizational structure is shown in Fig. 6.7.

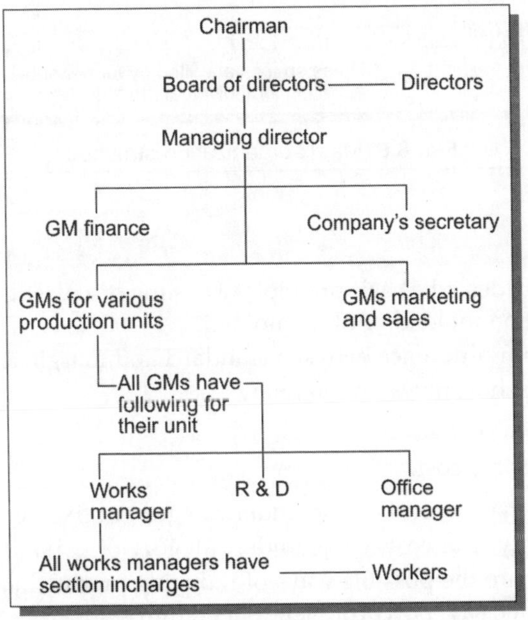

Fig. 6.7: Large company organization structure

Matrix Organizational Structure

Matrix organizational structure may be product based. There may be a set of activities such as:

- Research and development
- Production of specific product or its range
- Sales
- Finance

Each of product has a set of above mentioned activities. This concept of organizational structure is displayed with the help of Fig. 6.8.

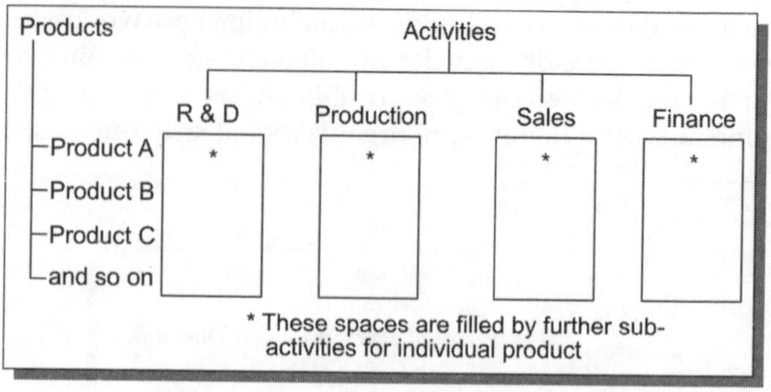

Fig. 6.8: Matrix organization structure

TEST

1. Briefly describe basic principle of costing of a plastics product, injection molded or blow molded.
2. Explain difference between standard and marginal cost.
3. Write short notes on following:
   - Fixed cost
   - Variable cost.
4. What type of costs come under category of fixed cost?
5. What type of costs come under category of variable cost?
6. What are the possible ways of reducing production cost of a product, say polypropylene sheet, film, etc.?
7. What are the possible ways of reducing production cost of an injection molded item?
8. How does a production in-charge, GM, MD of a production company may contribute to cost reduction of various products which company produces?
9. How does *Brand* image of a product affect its cost in market place?
10. What is break even point in costing of a product in relation to production quantities over a period of time?

ASSIGNMENT

Note: Tick mark in the box which is appropriate. Cross remaining

1. Which of the following comes under fixed cost of a production set up?

(a) Machine depreciation .... ☐
(b) Molds and tools maintenance .... ☐
(c) Salaries .... ☐
(d) Bonus .... ☐

2. Which of the following comes under variable cost?
   (a) Building .... ☐
   (b) Management overhead .... ☐
   (c) Out side maintenance cost .... ☐
   (d) Interest on money blocked in spares inventory .... ☐
3. Which of the following factors affects production cost of a thermoforming product?
   (a) Variety and grade of sheet material .... ☐
   (b) Sheet heating system .... ☐
   (c) Capacity of vacuum tank .... ☐
   (d) Value of air pressure .... ☐
   (e) Speed of machine .... ☐
4. Which of the following affects production cost of injection molded item?
   (a) Make of molding machine .... ☐
   (b) Day or night shift .... ☐
   (c) Number of cavities in mold .... ☐
   (d) Microprocessor controlled machine .... ☐
5. Which of the following is management organization system?
   (a) Hierarchical .... ☐
   (b) Flat organization .... ☐
   (c) Collective working .... ☐
6. While employing a person for an operator's job, which of the following is to be judged?
   (a) Physique .... ☐
   (b) Speaking power .... ☐
   (c) Aptitude .... ☐
   (d) Intelligence .... ☐
7. For *healthy* and profitable working in a production shop, which of the following is needed?
   (a) Playing of music .... ☐
   (b) Motivation .... ☐
   (c) Extreme rigidity in enforcing instructions .... ☐

8. Leadership in an industrial organization may mean which of the following?
   (a) Impose orders .... ☐
   (b) Ability to understand other's view point and then to guide and instruct subordinates .... ☐
   (c) To demonstrate to workmen that their supervisor is capable of doing all operations which workmen do .... ☐
9. Destiny of an industrial organization depends on which of the following?
   (a) Number of employees .... ☐
   (b) R & D facility .... ☐
   (c) Visionary board of directors .... ☐
10. Which of the following has high overheads?
   (a) Multiproduct producing company .... ☐
   (b) Cottage industry located in a town .... ☐
   (c) Multinational companies .... ☐

7 Practice–Plastics Processing

AUTOMATIC INJECTION MOLDING MACHINE

1. Idle run observation.
2. Study control knobs and computerized control console.
3. Study of safety precaution from machine operation instruction manual.
4. Study of injection unit, clamping unit and safety installation.
5. To study and well-understand procedure for starting an injection molding machine, study operating manual of machine.
6. Make sketches of machine plates, clamping system as exist in machine under study.
7. If mold is already set, observe and study mold function.
8. Operate machine in manual, semi-automatic and automatic mode.
9. Produce components, note down setting parameters and calculate production quantity per 8 hours shift.
10. Do quality checking and do SQC of at least 100 components.

MICROPROCESSOR CONTROLLED INJECTION MOLDING MACHINE

1. Study operation/instruction manual of machine to understand all safety precautions.
2. Study technical literature of microprocessor fitted on the machine.
3. Prepare a comparative notes on microprocessor and non-microprocessor machines.
4. Procedure of process parameter settings.

5. Setting of mold on machine
6. Molding with different plastics
7. Quality checking of product and preparing reports
8. Analysis of defects, causes and suggest remedies, prepare a write up.

EXTRUSION PROCESS (FOR BLOWN FILM, STRAIGHT FILMS AND SHEETS, PIPES AND TUBES)

1. Study of extrusion plant which includes extruders and ancillary equipment.
2. Study of safety precaution
3. Setting up processing parameter such as:
   - Extruder cylinder temperature at different zones
   - Screw speed
   - Nip roller speed
   - Winder speed
   - Blow ratio
   - Bubble cooling air control
   - Fixing of blown film die
   - Familiarizing with die parts and their function
   - Specification of machines to produce different sizes of blown films
   - Practice of die setting on machine
   - Practice of die setting, during running of machine, to achieve uniform thickness of film.

COMPRESSION AND TRANSFER MOLDING (ON SEMI-AUTOMATIC MACHINE)

1. Study of machine operation manual for safety precaution and procedure for operating the machine.
2. Study type of mold clamping on machine, procedures.
3. Setting of temperature of thermoset mold, curing time, clamping force and ejection strokes.
4. Analysis of product quality.
5. Identifying defects and their causes and writing down remedies.
6. Practice of doing production with thermosetting plastics.

AUTOMATIC BLOW MOLDING MACHINE

1. Study operating manual of machine.
2. Read and understand safety precaution and safety installation.
3. Machine setting procedure.
4. Fixing of mold on the machine.
5. Fixing of parison die.
6. Producing parison for blowing.
7. Machine operating practice with different mold.

THERMOFORMING MACHINE

1. Study instruction manuals of main machine and its ancillary equipment.
2. To note down operation sequence while operating the machine as per instruction given in manual.
3. Examine quality of products, make notes.
4. Note down actual defects, their causes and remedies.

For performing listed practicals, a good understanding of machine and its setting is necessary. Attempt is hereby made to provide opportunity for understanding through *How To Run* for various machine in following paragraph.

How to Run an Injection Molding Machine (Non-micro Controlled)?

A typical injection molding machine is chosen to describe how to run injection molding machine having conventional control panel as shown in Fig. 7.1.

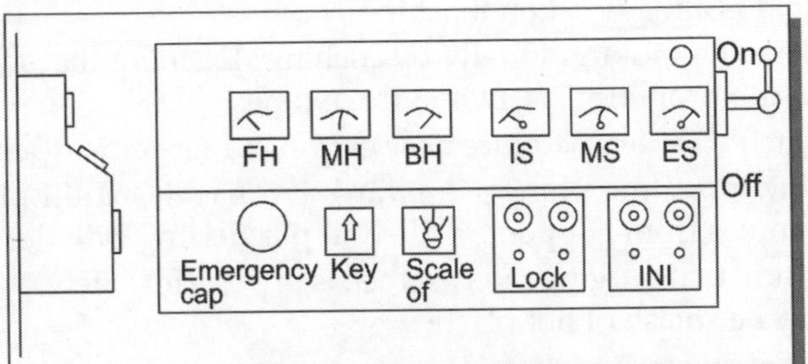

Fig. 7.1: Control panel

On right side of panel is a lever switch to ON/OFF power supply. On vertical face of panel are two sets of meters. On right hand side are timers for setting injection stay (IS), mold stay (MS) and ejection stay (ES). On left hand side are temperature setting of band heater in front (FH), middle heater (MH) and back heater (BH). On slant surface of control panel, there are three sections. Left hand section contains big red emergency STOP button. Then there is a START key. If key is in upright position, then semi-auto and auto commands remain isolated. Only inching can be done. Next section is for locking side. There are two push button switches. Green button is for mold opening and red is for closing. There are red and green indicator lights also. Third section is for injection side of machine. Green button on right hand side is for moving injection piston back. Red button is to move plunger towards left for injecting plastics in mold or for purging.

For operating this electro-mechanical machine, following steps may be adopted.

- Carry out check, if all three phases of power are there
- Check all safety devices
- Fill machine hopper with granules
- Cooling water is circulating through mold
- Put lever switch ON, red *Power ON* light glow
- Adjust required temperature on FH, MH and BH
- Wait till set temperature is achieved, indicated by small light on meters
- Set desired timing in timers IS, MS and ES
- Rotate selector switch knob to inching position
- In inching state, safety door may be opened
- Mold is already set on machine
- Close the mold gradually by inching. Examine that mold is closing properly. Try two to three times
- Watch if set temperature near each band heater is reached
- Now move the plunger forward and backward till plastics granules from hopper are filled in plasticizing cylinder
- Ensure that purging safety screen is in place for safety, as there may be splash of hot plastic
- Purge one or two shots
- Close the mold carefully by inching button

- Inject plastic in *one go* by pressing red button
- Give sometimes (say 5 seconds) to plunger remain in pushing condition of plastics
- Take plunger back
- Give mold stay time, say 4 seconds
- Open the mold. Examine molding
- Now put the key in its slot turn it in clockwise direction
- Bring selector switch on semi-automatic operation and close safety gate
- Press red button of mold close momentarily
- This will initiate full molding cycle
- After opening of mold and ejection of component. Take out four to five shots
- Bring selector switch on fully automatic operation
- Press red *mold close* button. Machine will start running fully automatic
- Collect about 10 shots. If mold has four cavities then number of molded components would be forty
- Pick up five components of each cavity and prepare quality checking report or observation sheet. Proforma of observation sheet may be as per instruction of supervisor or in-charge.

How to Run a Microprocess or Controlled Injection Molding Machine?

Before attempting to run the machine, it is necessary to understand working principle, design and construction of microprocessor controller.

Microprocessor is a device in which sequence of operations, desired parameters for variables such as time, temperature, speed, pressure are loaded. There are facilities for operator or setter to intervene. Loaded parameters and interventions are *Inputs* to microprocessors. Another channel of input is machine itself. It has got a number of sensors and limit switches which provide input signal to microprocessor. It in turns analyses the data electronically and develops a set out put signals for machine to work accordingly. Almost all modern injection molding machines are equipped with PID (proportional integral derivatives) and PLC (programmable logic control). All these sophisticated equipment and various sensors of machine work in unison to create very accurate machine control outputs.

Please note that there are many reputed companies in various countries who make microprocessor control panel suitable for various plastics processing machineries.

Before going for practicals work, study technical literature of microprocessor controller which is fitted on that particular machine on which practical are to be performed.

For the sake of providing a general idea, a sketch of a microprocessor controller is given in Fig. 7.2.

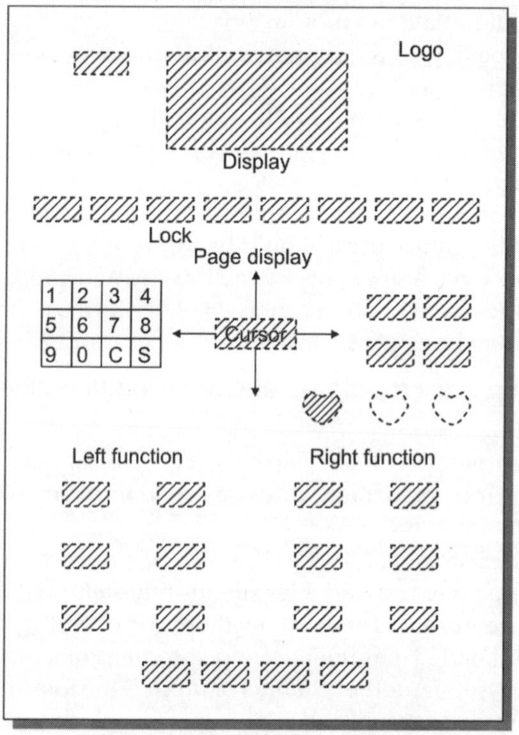

Fig. 7.2: PLC digital control panel

On the top, there is a digital display screen. Blow the screen are a number of buttons to select a large display. Since all information, functions and controls cannot be displayed at one time on screen, therefore, all the functions are divided into individual screen. If setters want to view details of locking side then he would press or touch button *lock*. Like personal computer or laptop, a page *locking side* may pop-up where details for choosing and setting may be displayed such as locking speed, first stop distance, second stop (if required) distance, final stop, locking force, ejector operation sequences and speed. Similarly, there are many details of units such as injection, feed, temperature, etc. for this another page may be brought on screen by pressing or touching appropriate buttons.

Please note that there may be more than one mold which may be run on the machine. Each mold is given a code number and a set of particular setting of parameter which are required and established from previous run. So there is a provision of calling a *mold* page where a particular mold may be selected. On selection, all relevant settings and parameters are displayed. That particular mold is already once set on machine. In case, setter desires to do some fine correction or changes in parameter, can do so with the help of left or right function and numeric keys (touch buttons). Similarly, temperature page may also be brought on screen for checking various set parameters. If necessary, *fine tuning* can also be done.

Common functions like settings, manual adjustment, auto-operation and *home page* may also be brought on screen.

Users of PCs and laptop may not have much difficulty in completing the setting even for a new mold and save the setting.

Facility of *password* is normally provided so that unauthorized persons cannot activate microprocessor controllers.

Sequence of action may be as follows for operating microprocessor-controlled machine.

- Put main switch of power ON.
- Activate control console.
- Call page *temperature*, put the heating ON.
- Call page *mold*, put water circulation ON through built in mold temperature control unit.
- Call page *hydraulics* starts the pump by touching appropriate button on this page. Allow pump to run so that temperature of oil get stabilized before machine is started for production.
- When plasticizing cylinder has attained temperature which will be indicated by light indicator and a sound alarms.
- Call page *mold* again and enter mold code no. On entering mold code, all related parameters would be displayed for that particular mold.
- Touch OK button

- Make changes in parameter, if desired. In case it is desired that from now onwards mold travel for locking should stop for the period till *continue command* is reached from core operating system.
- Call page *operation*, touch key of *inching* or *manual*. Do purging once or twice now run full molding cycle.
- Check if molding quality is OK.
- If yes, touch semi-auto-key and touch *start* key. Full molding cycle would take place.
- Once satisfied with machine mold operation and quality of component, go ahead.
- Touch *fully auto* button, give *go* command, machine will start molding in fully automatic mode.
- In case of any malfunction, safety system would stop the machine and would give light and sound alarm.
- Machine will also stop automatically, if safety door is opened or any safety guard is removed while machine is running.

Photo 7.1 shows an injection molding machine fitted with microprocessor control panel.

Photo 7.1: Microprocessor control panel

Courtesy:University Polytechnic, AMU

How to Run a Fully Automatic Blow Molding Machine?

Like fully automatic injection molding machine, blow molding machine is also equipped with PID and PLC which take inputs from machine sensors and microprocessor. They also provide output to microprocessor, all the three work in unison.

Microprocessor control panel have different configuration of keys, pages and logic as compared to that of injection molding machine.

Please note that sequence of operations given below is a typical one. Blow molding machine on which practical is to be performed may be having microprocessor control console from a particular make. It may have a specific model. All these may have a different set of controls. It is, therefore, highly recommended that instruction manual for machine and console may be studied. However, given sequence of action may give a fair idea as to what controls may be there and are they configured.

In case of automatic blow molding, following pages may be *called*. This means that pages come on screen of which button is touched.

- Menu
- Carriage
- Blow pin
- Pneumatics
- Shot monitor
- Temperature
- General
- Miscellancous (misc)

Let blow molding machine available for practical be a accumulator type non-continuous blowing machine and a bottle mold is already loaded. Sequence of action may be as follows:

- Check power and high pressure air supply.
- Put ON power to microprocessor control by touching a button. Screen lights up.
- Touch mold page *call* button, mold page appears on screen.
- Select mold code for the mold already loaded on machine .
- All parameters such as melt temperature, force of accumulator plunger on plastics, downward speed of parison, etc. to be checked, if found ok then ok button is to be pressed or the cursor is to be brought to a particular value of setting for making any correction by rewriting the value with the help of numeric pad and then enter.
- *Call* operation status page to select mode of operation, suppose semi-automatic mode is selected.
- Touch start button.
- Parison start flowing out of die opening in the form of tube.
- Mold in open condition on carriage, moves towards right to its set position so that parison is coming down in such a way that axis of parison and cavity coincides.

- Parison reaches its down limit and accumulator plunger stops pressing the plasticized plastics.
- Mold immediately closes.
- High pressure air enters enclosed parison and blows it to take shape of mold cavity profile. This happens very fast.
- Pressure pipe is withdrawn downwards.
- Carriage moves to left and mold opens.
- Blow bottle is ejected by means of air blast or in combination with ejector pins.
- Take out few number of bottles of polypropylene.
- Switch off heating only and not cooling water circulation.
- Examine quality of bottle for various quality points and prepare an observation sheets.

How to Operate Vacuum Forming Machine?

Let there be manually operated vacuum forming machine for carrying out practical. Basically, there are two stations. One for heating the sheets and other the vacuum forming station. There is a sliding frame on which piece of sheet is placed. Frame is slid under heating hood. After proper heating, frame is slid over cavity which is connected to vacuum tank.

Sequence of operations would be as follows:

- Cut a piece from polystyrene sheet 0.35 mm thick and width and length to suit frame.
- Place the sheet on frame.
- Push the frame under heating hood in which heating elements are already hot.
- Give few seconds, say 20 to 45 seconds for sheet to become soft enough for taking shape of cavity under vacuum.
- Push the frame on cavity station and clamp it to the face of die, which creates a seal between sheet surface and cavity block face.
- Put the vacuum switch lever to connect cavity to vacuum tank which instantly sucks the sheet into cavity. Hence, sheet takes shape of cavity and a component is formed.
- Open the clamp, remove the sheet with formed component.
- Place sheet over a trimming die station where component is to be trimmed down.
- Close safety gate.

- By pushing a button, activate trimming punch. It comes down and trims the component down. Scrap sheet remains on the face of trimming die.
- Open the safety enclosure and remove the scrap by air blast.

STUDY OF MOLDS

1. Study at least two injection molds.
2. Prepare sketches and name the parts.
3. Describe, if mold is a 2-plate or 3-plate , type of runners , type of gates, ejector system.
4. Prepare a list of possible defects in molds. Mark defects which are found in the mold under study.

An observation sheet is provided here as a possible format.

Mold Condition
Observation Sheet

Name of observer: ....
Date of observation: ....
Component: ....
Mold code number: ....

Type of mold:	Two plate ....
	Three-plate ....
Type of runner	Full round, half ....
Gate design	Direct ....
	Fan ....
	Banana ....
	Sub-marine..........................
Cavities layout	Balanced ....
Condition of guide pins	Good ....
	Poor ....
Type of ejector	Pin ....
	Sleeve ....
Provision of cooling	Yes ....
	No ....
Code No provided	Yes ....
	No ....
General up keep	Satisfactory ....
	Not satisfactory ....

Remarks : ....

Signature

FINISHING OF PLASTICS COMPONENTS

Many molds are still in use where de-gating of components from runner is manually done. Finish of de-gated portion should be such that it gives feeling of smoothness, when touched. Moreover, the finish should be such that it becomes difficult to locate the position of gate on component. Manual de-gating is normally done by using a sharp knife. De-gater tries to cut gate very close to component surface. It is very difficult to maintain uniformity in cutting. Sometimes cutting takes place a little away from component surface and sometimes a little portion of component is also cut together with gate. In case of round component, small trace of protrusion remains in spite of very close de-gating. This happens when diameter of component is less and width of gate is more. This problem is minimized to a great extent by using a curved cutting edge which matches surface arc of component.

Often a jig is used for de-gating. Arrangement generally is that there is provision of a stopper to suit shape of component. Cutter has a fixed position of movement with respect to stopper. Consequently, almost uniform de-gating takes place consistently.

On many occasions, buffing is done, in case of thermoset molding, if there are flashes or small protrusions. Nowadays generally molds and molding process is so accurate that there is hardly any need of finishing operation.

Surface of plastics sheets is treated so that printing inks adhere properly to the surface. Treatment is that surface is exposed to corona radiation which makes surface of sheet suitable for printing process. Injection-molded articles are sometimes printed by manual screen printing or by automatic rubber pad transfer printing method. In this technique of finishing, components produced from polystyrene or ABS hardly need any surface preparation. Components produced from HDPE, PP, PET, etc. may need surface treatment.

Finishing also includes spray painting of large components like TV cabinet, front panel, computer mouse, some autoparts, etc.

There are many methods of treating surface of plastics products like injection-molded items, extruded films and sheets. Pipes, profiles and blow-molded items. Some of these methods are given as follows.

- Chromic acid etching
- Corona discharge for polyolefins, ABS, polystyrene, acital, etc.
- Flame treatment
- Iodine treatment
- Plasma treatment
- Primers
- Surface roughness
- Thermal treatment
- UV exposure

For studying the details of above, it is recommended that the following website may be visited.

www.specialchem4adhesives.com
(search adhesive guide)

Another technique of providing attractive finish is to electroplate ABS components which are molded from plating grade ABS. It provides bright shiny surface. Many automobile components, handles, caps, etc. are electroplated with chromium.

It is a well-known fact that metallic components can be electroplated as metals are good conductor of electricity. ABS plastics cannot be electroplated in its natural form as it is not electrically conductive. But ABS surface can be made conductive by treating it with electroless process. It means etching of surface with etching solution based on chromic acid. Electroless copper deposition is possible. Once a very thin layer of copper is deposited then it becomes electrically conductive. Now normal electroplating may be carried out.

TEST

1. What do you understand by doing practicals related to plastics technology?
2. Which are most common machineries used for converting plastics materials into product?
3. How would you learn to operate a machine, say injection molding machine, if you have never operated?
4. Describe usefulness of operating manuals which are generally supplied with the machine?

5. List out safety precautions which should be observed at the start and during production. Briefly describe each.
6. Why is it very important that safety precautions be observed? What may be the consequences of negligence towards safety measures?
7. What preparations are necessary for starting production of lay flat tubing plant?
8. Describe step by step actions to do thermosetting molding on manually controlled press.
9. Describe step by step actions to start a manually controlled injection molding machine.
10. Describe step by step actions to start and do production of PET bottles by using pre-forms.
11. Explain why does microprocessor-controlled machine give better results in terms of product quality and productivity?
12. Write full form of following acronyms:

    PID

    PLC

    PET
13. Describe step by step actions to start an extruder plant for covering metallic wire with PVC.
14. Briefly describe various types of blow molding machines.
15. Briefly explain how does sequence of operations achieved in an injection molding machine, specially hydro-mechanical machine.

ASSIGNMENT

Note: Tick mark in box for appropriate options. Cross remaining

1. Which one of the following options is correct to describe difference between two and three plate mold?
   (a) Two-plate mold has only two plates in its construction .... ☐
   (b) There is no floating plate on injection side .... ☐
   (c) In a three-plate mold, there is a floating plate in mold half of injection side .... ☐
2. Injection molds generally have two or four guide pins. One is kept bigger in diameter as compared to other. Which of the following is appropriate reason for that?

(a) To hold all plates in place .... ☐
(b) To make the mold strong .... ☐
(c) To ensure correct matching of two halves of mold .... ☐

3. In a sheet extrusion die, there are a number of bolts with locking nuts along the width of die. Which of the following described purposes is appropriate?
   (a) To make die strong .... ☐
   (b) To hold die lip .... ☐
   (c) To adjust gap between lower die plate and upper lip ... ☐
4. In blow molding process, wall thickness of product is adjusted by which of the following
   (a) Speed of parison formation .... ☐
   (b) Varying temperature of extrusion head .... ☐
   (c) Adjusting gap between die ring and core .... ☐
5. Which of the following processes is suitable for producing FRP corrugated sheets?
   (a) Extrusion .... ☐
   (b) Compression molding .... ☐
   (c) Manual lay of fiber glass strands, spraying epoxy with chemicals, pressed and heated .... ☐
6. Which of the following parameters is important for blow molding machine set up?
   (a) Accurate setting of timers .... ☐
   (b) Accurate setting of temperature .... ☐
   (c) Accurate setting of speed of mold travel .... ☐
7. Which of the following settings can be done on microprocessor based control panel?
   (a) Travel speed of injection unit .... ☐
   (b) Injection speed .... ☐
   (c) Injection pressure .... ☐
   (d) Locking force .... ☐
8. Which of the following processes is much complicated? Variety of components can be produced and it is much in use
   (a) Rotational molding .... ☐
   (b) Thermoset molding .... ☐
   (c) Injection molding .... ☐
9. Which of the following processes yields higher percentage of scrap in batch run?
   (a) Extrusion of lay flat tubing .... ☐
   (b) Extrusion of sheets .... ☐

(d) Extrusion of pipes .... □
(e) Injection molding with conventional runner type mold .... □
(f) Injection molding with sprueless mold .... □
PET bottle blowing .... □

10. Which of the following checking may be done before start of injection molding machine?
(a) Mold halves are properly aligned .... □
(b) Molds are painted and identifying code given .... □
(c) All safety systems of machine and mold are working ... □

8 Practical–Plastics Testing

INTRODUCTION

A variety of products are made from plastics. These consist of household items, toys, durables and technical items such as gears, auto accessories, medical gadgets and many others. All such items vary in quality standard and pricing. Certain items such as medical gadgets, electrical and electronic components are to be made of high quality standard to meet demand of customers and fulfill statutory obligations.

To produce good quality standard product, it is necessary to select plastics of appropriate properties. This is in addition to proper control over molding, extrusion or whatever process is adopted for converting plastics materials into products.

In practice, processors generally do not maintain a test lab for testing plastics materials. They mostly depend on properties declared by plastic materials manufacturers. Big corporate companies producing large quantities of plastics parts and components from various types and grades of plastics may have in-house testing facilities. Plastics testing facilities are available in our country also.

There are few concerns who carry out tests, determine properties of sample provided by customers. Renowned plastics teaching and training institutes of the country generally have well-equipped testing laboratories. These institutions may be approached for getting plastics materials and products tested. List of few testing sources are given below.

www.polymer-testing.com
www.international-equipment.com
www.tiniusolen.com

www.testlopedia.com

www.idea.com/articles/plastic_test_standards.asp

www.wikipedia.org

www.google.com/Shriram Institute (search for polymer testing facility, India)

VARIOUS TESTS AND TESTING EQUIPMENT

There are national and international agencies who have standardized various test parameters and methods. Advantage of this standardization is that products from various sources would almost have the same performance provided producers have followed national and/or international standards.

For on-going consistency in performance of a product, it is necessary that plastics granules/powder/resin manufacturers and processors may follow same standard norms:

Name of few standard organizations:

- BIS – Bureau of Indian Standards
- ASTM – American Standards Test Methods
- ISO – International Organization for Standardisation
- EN – European Standards
- DIN – Deutsches Institut Für Normung
- BS – British Standards
- JIS – Japanese Standards Associations

This chapter provides basic information about carrying out practical tests. Most common tests generally performed in testing labs are:

- Melt flow index
- Specific gravity
- Vicat softening point
- Abrasion test
- Flexural strength
- Tensile strength
- Izod impact
- Tear strength
- Hardness
- Burning
- Permeability

- Weather effect
- And many other

Vicat softening point testing equipment is shown in Photo 8.1.

Photo 8.1: Softening point testing machine *(Courtesy: University Polytechnic, AMU)*

All above tests are designated by code numbers and/or alphabets. Some typical examples are indicated in Table 8.1

Table 8.1: ASTM and ISO test numbers for some properties

S. no.	*Test*	*BIS*	*ASTM*	*ISO*
1.	Melt flow index	–	D 1238	1133
2.	Specific gravity	–	D 792	1183
3.	Flexural strength	–	D 790	178
4.	Tensile strength	–	D 63	527
5.	Izod impact	–	D 256	180
6.	Falling dart impact (film)	–	D 1709	178
7.	Hardness	–	D 2240	-
8.	Coefficient of friction	–	D 1894	-
9.	Weathering-QUV	–	D 4329	4892

Further, a long list of plastics related IS tests is given at following web address

www.bis.org.in/sf/pow1.asp?d=PC&c=012&x=Plastics+Sectional+Committee

It is recommended that this site may be visited.

Melt Flow Index

Melt flow index is defined as the weight of polymer which comes out of an orifice (capillary) of specified diameter and length at a definite polymer temperature, pushed by a specified weight in 10 minutes.

Instrument used for testing melt flow index is called melt flow rater. Various makers of this instrument give trade names of their choices.

Figure 8.1 gives idea of construction of a typical melt flow rater.

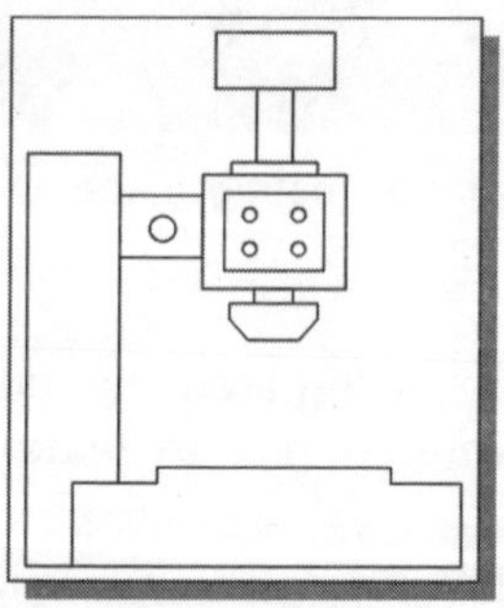

Fig. 8.1: Melt flow rater

Photo 8.2 of melt flow testing equipment (melt flow rater) is also given.

IS and ASTM provide detailed specification of melt flow rater and operating method. Modern testing equipments are fitted with computerized recording system and indicate fixed as well as set parameters.

Note: It is very important to read instruction manual of testing equipment before handling it. Safety precautions and control points may vary from make to make and model to model.

Photo. 8.2: Melt flow rater *(Courtesy: University Polytechnic, AMU)*

Following steps may be followed to operate MFR (melt flow rate):

- Remove push rod from MFR cylinder
- Take approximate quantity of polymer granules/powder/resins
- Pour polymer in cylinder up to specified height
- Place the push rod in cylinder, over granules
- Put ON heating
- When set temperature is achieved, it would be indicated
- Put collector dish on the platform under orifice
- Set time for which polymer is to extruded (come out of orifice)
- Now cut out hung *thread* or *strands* from orifice face by means of attached cool knife
- Push start button
- On completion of time, polymer strand would get stop automatically or manually
- All test parameters would be indicated on display screen or panel
- Melt flow index may directly be read
- If facility is there in MFR then a printout may be taken.

Figure 8.1 shows a typical melt flow rater

Specific Gravity Measurement

Figure 8.2 shows schematic sketch of a specific gravity determining equipment. (1) is the main frame of instrument/machine. (2) is control panel/console to enter test parameters and to record reading from LCD display window. (3) is a lever coming out of frame. Balance of wire construction is hung from lever (3). (5) is a fixed dish on which sample is placed. (6) is a beaker having distilled water in which sample is hung. Beaker (6) is placed over a stand (7). From underneath, the stand passes the wire frame of balance in such a way that it does not touch the stand.

This equipment is suitable for determining specific gravity of those plastics which are heavier than water.

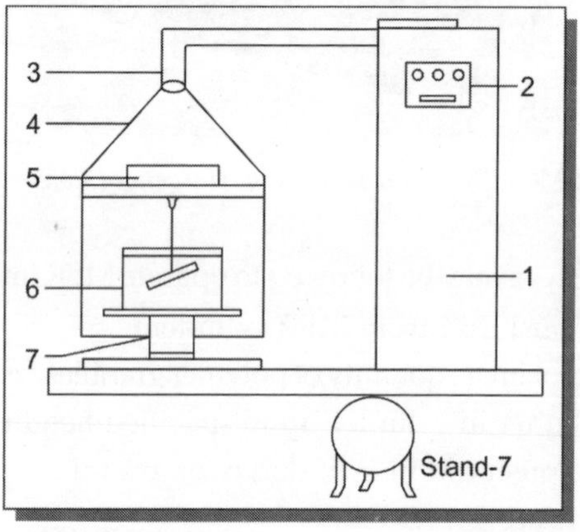

Fig. 8.2: Specific gravity measurement

Procedure for Testing

Procedure for determining specific gravity would be as follows:

- Place the beaker with distilled water on stand
- Place sample piece of plastics on fixed dish (5)
- Press button on control panel. Weight of sample in air would be indicated and machine would electronically record it for further process/auto-calculation
- Press freeze button
- Remove specimen from dish (5)
- Hang it from hanger hole in such a way that specimen remains under water. Delicately shake the sample so that no air bubble remains clinged to specimen surface.

- Push test button. Weight of specimen in water would be indicated and also specific gravity would be directly indicated.

Izod Impact Test

Figure 8.3 is a schematic display of izod impact tester. Names of various parts are mentioned below

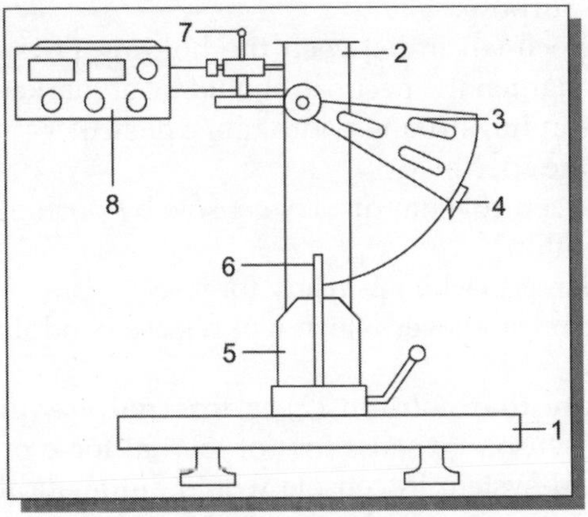

Fig. 8.3: Izod notch impact. 1. Base, 2. Vertical column, 3. Impact pendulum, 4. Impact blade, 5. Specimen holding vice, 6. Specimen, 7. Pendulum release lever, 8. Control cum indicator console.

Procedure for Testing

Please note that to carry out izod impact test, specimen is to be prepared as per dimensions shown in Fig. 8.4.

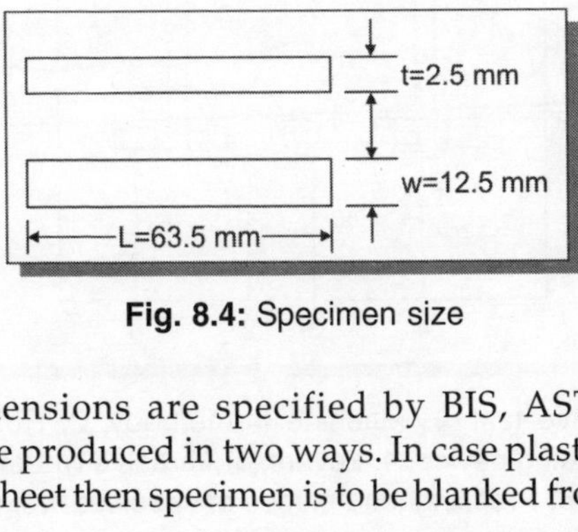

Fig. 8.4: Specimen size

These dimensions are specified by BIS, ASTM, ISO, etc. Specimens are produced in two ways. In case plastics material to be tested is a sheet then specimen is to be blanked from it by means of a blanking tool. If polymer is in the form of granules, powder and resins then the specimen is to be molded by injection or compression molding machine.

Procedure of Testing

Procedure for testing izod impact may be as follows:

- Lock pendulum at its top position. There is lock/release lever for this purpose.
- Place specimen in between the holding jaws of vice and slightly tighten it. Specimen should be at marked location so that it is in full striking width range of striker
- Clamp the specimen
- Activate control cum display console by pushing the appropriate button
- Now testing machine is ready for test
- Operate release lever which will release pendulum to swing freely.

Please note that column (2) is internally equipped with appropriate sensors to sense loss of inertial force of pendulum. Computerized system in console would automatically calculate and display izod impact value in ft-lb per inch square or in metric unit, metre-kg/cm^2.

Abrasion Test

Figure 8.5 shows schematic construction of a abrasion testing machine. It consists of following parts:

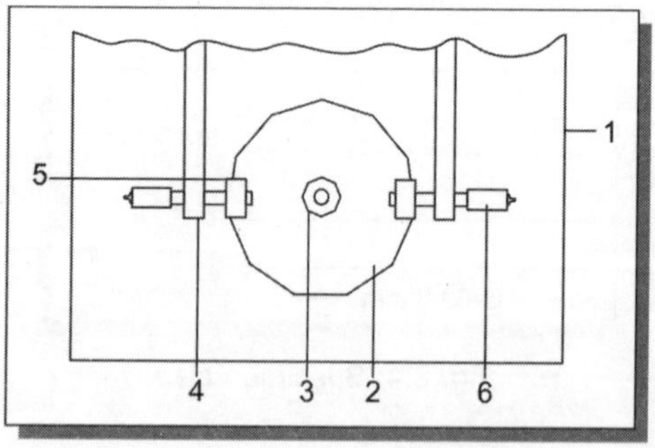

Fig. 8.5: Abrasive test. 1. Machine frame/body, 2. Rotating platform, 3. Specimen holding device, 4. Levers, 5. Abrasive wheels with specified fine roughness, 6. Pressing load pieces of specified weight, changeable

Note : Construction, setting, control and indication may vary from make to make of machine. It is, therefore, necessary to study instruction manual of machine before operating it. Specimen under test has to be a disc with a hole in the center for clamping on rotating platform.

Dimensions of the specimen are specified by BIS, ASTM, BS, EU, ISO, etc. and they are prepared accordingly.

Testing Procedure

- Fix specimen disc, with central hole, on rotating platform
- Place abrasive wheels (attached to levers) delicately over the specimen
- Place loads of specified weight on both side levers
- Set r.p.m and time of operation on control cum indication console
- Start vacuum system which sucks particles becoming loose due to abrasion
- Start platform to rotate.

Please note that weight loss would be indicated after specified set time, say two hours or four hours, etc.

Quantity of abrasion of a polymer surface is indicated by the loss of weight it suffers due to abrasion process.

Universal Testing Machine

Photo 8.3 shows a universal testing machine and Fig. 8.6 shows its schematic construction.

If a specimen of a polymer is stretched gradually then following happens to specimen:

- Stretching only in elastic limit. This means that specimen is stretched only so much that it attains original length, when released, without any permanent deformation. Depending on type and grade of plastics, amount of stretch may be very small.
- If specimen is further stretched, a situation would arise where further stretching will not take place and stretched portion will not attain its original length. This is tensile strength at yield.
- On increasing stretching force, a point would come when specimen would break. The amount of force applied per unit cross-section.

Description of Machine (Fig. 8.6)

Values recording system is fully automatic. At the end of testing, data sheet and graph is obtainable, depending on the command given at control console at the time of starting the machine. Print-out of data and graphs may be taken out afterwards, even if the command is not given in the beginning.

1. It is the base of machine on which two columns are mounted on the sides. A stationary griper is also fitted to grip specimen to be tested
2. Side columns
3. Stationary gripper
4. Movable grippers to hold and pull the specimen
5. Hydraulic jack mechanism
6. Control console
7. Display monitor screen
8. Printer

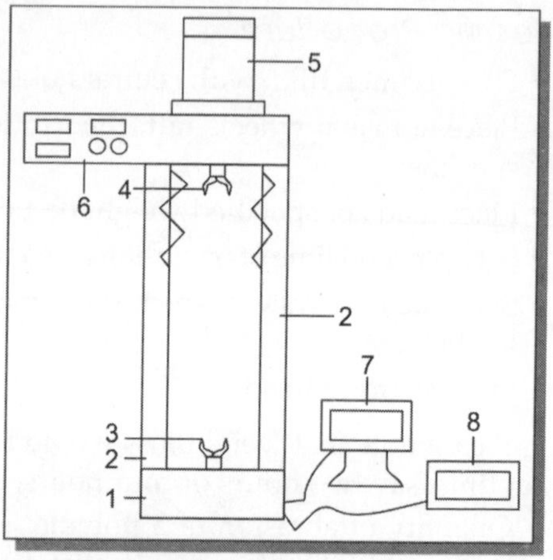

Fig. 8.6: Universal testing machine

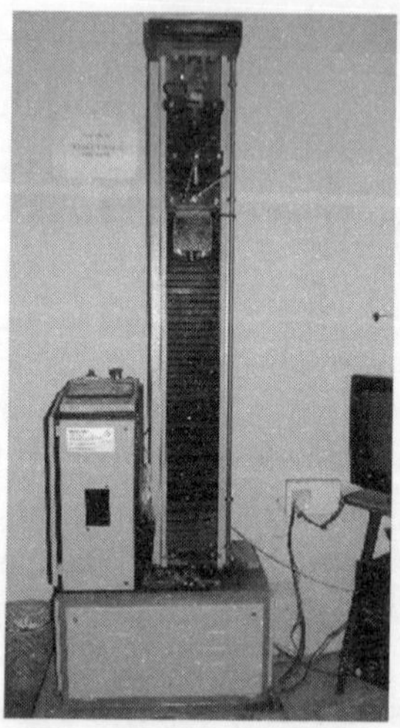

Photo 8.3: Universal testing machine *(Courtesy: University Polytechnic, AMU)*

Tensile Strength Test

The amount of force applied per unit cross-section area is tensile strength at break.

Figure 8.7 shows a typical shape of specimen for carrying out tensile test. Tensile test as per ASTM D 638 and ISO 527 may be carried out for obtaining:

- Tensile strength at yield
- Tensile strength at break
- Tensile modulus
- Tensile strain
- Percentage of elongation at yield
- Percentage of elongation at break

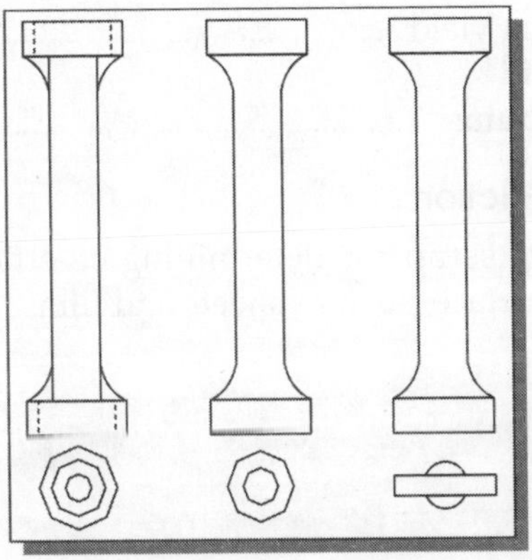

Fig. 8.7: Specimen

A separate strain gauge is attached to machine to test specimen as shown in Fig. 8.8.

It is equipped with *stretch* sensors to sense on going variation in *l*. Sensor is electrically/electronically connected to microprocessor. During test, continuous flow of data from applied load (pull) sensor and strain sensor enable the system to draw a graph on monitor screen.

After the test is over, results are indicated in following units:

- Tensile strength at yield .... psi or kg/cm^2
- Tensile strength at break .... psi or kg/cm^2
- Elongation a break .... %age

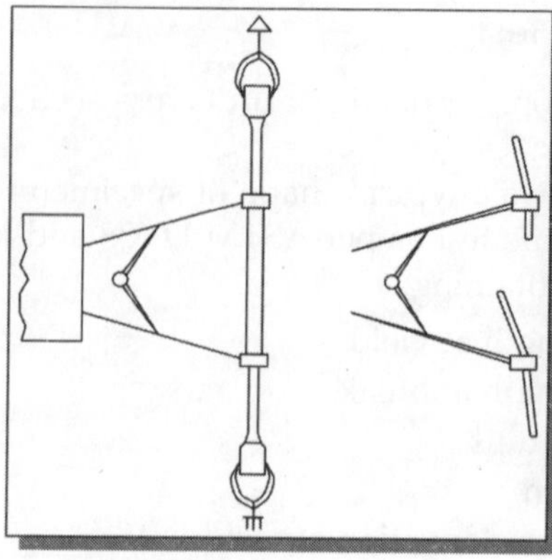

Fig. 8.8: Strain gauge

- Elongation at yield .... %age
- Tensile strain
- Tensile modulus .... 10^3 psi

Coefficient of Friction

Photo 8.4 shows setup for determining co-efficient of friction plastics article surface, such as sheets and films.

Photo 8.4: Co-efficient of friction tester *(Courtesy: University Polytechnic AMU)*

TESTING OF PIPES

A variety of pipes are produced to cater to general usage such as flexible pipes for conveying or siphoning water. Pipes of large diameter such as 100, 150 or 200 mm are produced to handle liquid under pressure. It is, therefore, necessary to know as to how much pressure a particular brand and type of pipe can safety stand.

Manufacturers of pipes get the sample of their product tested for the value of pressure in kgf per centimeter square which can be safely applied. Based on the result of test, they declare to the customers verbally and through catalogue, safe valve of pressure for which pipes may be used. Many customers get the specimen tested before purchasing bulk quantities, specially when *fail safe* strength pipes are required.

Plastics pipes are also tested for inside continuous pressure for a long time. There may be various time period targets which may even be extended to years to safely conclude that plastic pipes would not fail (creep or burst) during its service life.

There are many factors for which pipes are tested, few of them are mentioned below.

- Circumferential stress in pressurized pipes
- Temperature pressure relationship
- Friction loss characteristics
- Long-term strength

A significant reason of failure of a pipe is lack of circumferential strength against stress development due to pressure inside. Some of the test methods are generally adopted to test plastics pipes, specially PVC, CPVC pipes. Specifications are as follow:

- ASTM D 1598 — Long-term strength test
- IS 4984 — Hydrostatic pressure test method
- IS 4985 — Creep rupture test
- ASTM D 1785-06 — Standard specification for polyvinyl chloride (PVC)

For testing a pipe, a sample piece, holding jigs and a programmable testing equipment is needed. Figure 8.9 shows specimen holding jigs.

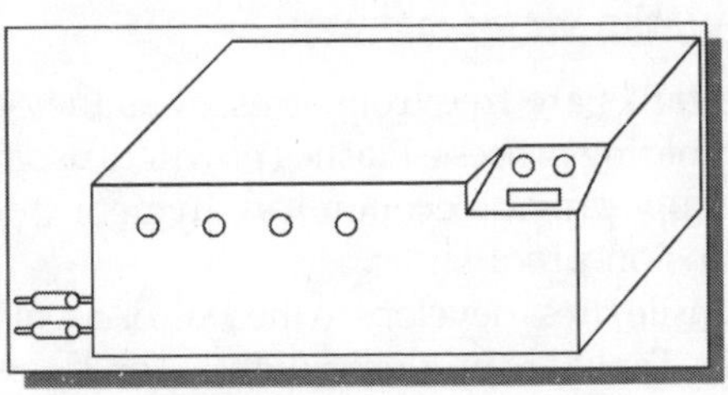

Fig. 8.9: Pipe testing machine

Jig has an arrangement to cap both the ends of specimen (piece of pipes of suitable length) firmly. There is an arrangement to connect the jig with machine and the air to escape, when pressure creating liquid, normally water enters the jig.

Basic construction of a typical testing machine is shown in Fig 8.10.

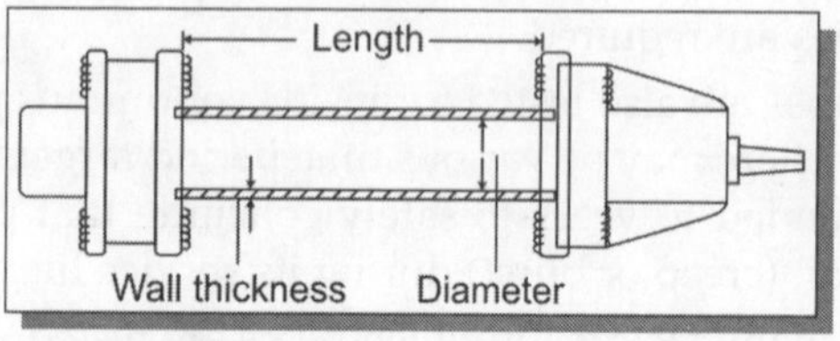

Fig. 8.10: Pipe testing jig

It is a computerized PLC based hydrostatic pressure testing machine with a number of testing parameters such as slow increase and decrease of pressure, test time and number of cycles. Machine is provided with a USB (universal serial bus) to connect to PC. A software may be used to program and implement testing. A printer may be attached to record data till failure occurs.

It is recommended that following websites may be visited as good amount of information may be found. Even the images of testing equipment and accessories may be seen

www.plasticpipe.org/pdf/chapter05

www.lascofittings.com/supportcenter pressureratingplastics.asp

www.astm.org/standards/D1785htm

www.internationalequipments.com/comp_hydrostatic.jpg

www.coleparmer.com/techlibrary/article/740

TESTING OF WATER STORAGE TANK

Water storage tanks are free from stresses as they are produced by rotational molding process. Plastics powders used for rotational molding generally provide economical, hygienic, odorless, leak-proof and corrosion-proof tanks.

Maximum tensile stress develops in the wall of circular tanks, when it is full of water. Tanks are produced with 1, 2 or 3 layers of different or similar types of plastics to make it mechanically tough and strong.

Destructive testing of water storage tank is rarely done. This includes dart impact test and weight drop test. Outside circumference of tank on the top, mid height and near bottom is accurately measured. Tank is then completely filled with water and circumference is measured once again. Increase in circumference may not be more than 0.2 to 0.5 % of diameter. Further, total body of tank should be strong enough so that it does not collapse, if subjected to weight load on top. Say, for a 500 liters of tank, approximately 80 kg if placed on the lid. Sometimes in transportation, tanks are subjected to point pressure from outside. This tends to deform the surface of tanks, specially wall. In case deformations do take place then tanks should attain its original shape back once outside force is removed. This can be tried in a jig by which a known force can be applied.

For specified and accurate testing methods IS 12701–1996 and / or ASTM D 1998–06 may be applied.

TESTING OF FILMS AND SHEETS

Plastic films and sheets may be tested for its tear strength, impact strength, haze, light transmittance, permeability to oxygen, burning rate, shore hardness, etc.

Sheets are mostly used for wrapping, laying, shelter making, etc. During use, films and sheets may be subjected to tearing force. Most of us must have noticed that tearing a piece of a film or sheet is rather difficult if there is no cut at the edge. Sheet may be torn much easily if a triangular wedge type small cut is made at the edge of specimen. Therefore, tear strength may be determined in two ways:

1. Without any cut
2. With cut

Results of tear strength are expressed in kgf or Newton.

Tear strength tester consists of specimen holding device, impact giving system where speed and force of impact may be altered. There may also be pendulum type impacting device. Display of reading may be analog or digital, depending upon type of machine.

Majority of machines are equipped with microprocessor, PLC and computer screen display.

Procedure for carrying out tear strength test may be some what different for various design and construction of machines. Basically, almost all manufacturers of testing machines aim at following the test procedure prescribed by Bureau of Indian Standards and ASTM.

Following standards may be studied for obtaining more information.

ASTM D 2261

- Prepare a specimen by cutting out a piece of films or sheets as per shape and dimension given in Fig. 8.11

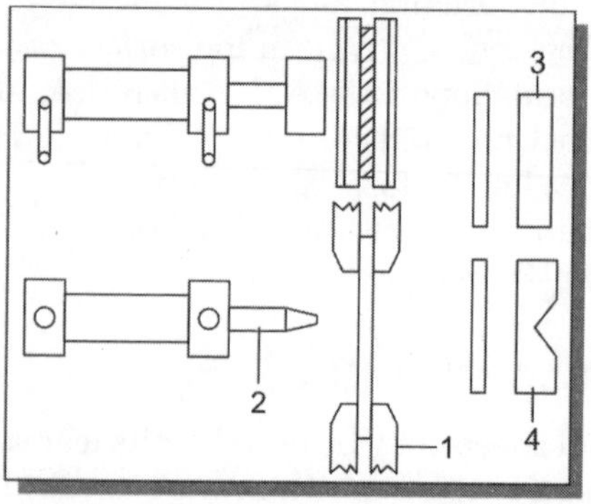

Fig. 8.11: Tear specimen and testing

Hold the specimen in between the jaws of holding vices (1).

- Set speed of pusher (2)
- Set maximum force limit
- Activate system, including microprocessor display screen and printer, if needed.
- Pusher (2) advances and touches the held specimen. Further, advance of pusher (2) meets with resistance due to resisting force generated by elongation of specimen.
- Distance moved by pusher and increase in force of pusher is kept on recording automatically and a graph is generated on display screen.
- As soon as specimen tears, motion of pusher stops.
- Tearing strength is in kgf at tear point as shown in Fig. 8.13.

Another instrument for testing tear strength is based on generation of striking force by means of pendulum. Fig. 8.12 shows basic construction of pendulum type tear strength testing machine.

Figure 8.12, (1) is specimen holding device. (2) is striker, (3) is a calibrated scale which displays potential energy in lbs-ft units of

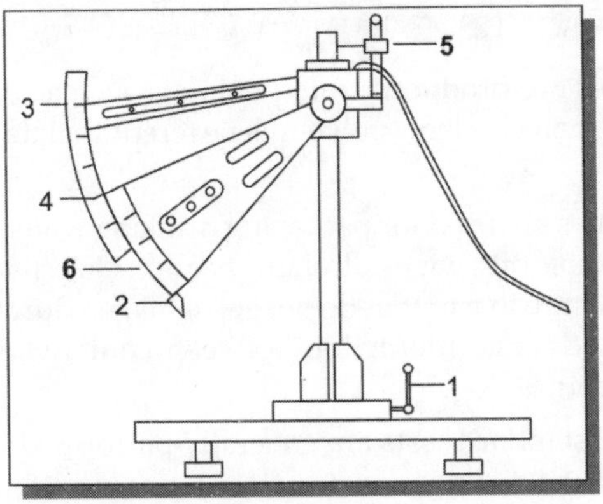

Fig. 8.12: Pendulum type testing machine

striker (2). When pendulum (6) is released by means of handle (5) then striker (having a definite weight) swings down. When striker reaches a perfect horizontal position, then potential energy takes a forms of inertial force in lbs by which striker strikes the specimen. Position of striker is varied to obtain higher or lower values of striking forces. By trial and error, correct setting of striker is obtained just enough to tear the specimen.

Please note that scale (3) may also be having another calibration in terms of striking force. Both the calibrations may be provided on the same plate which is in an arc shape and permanently fixed to system column (7).

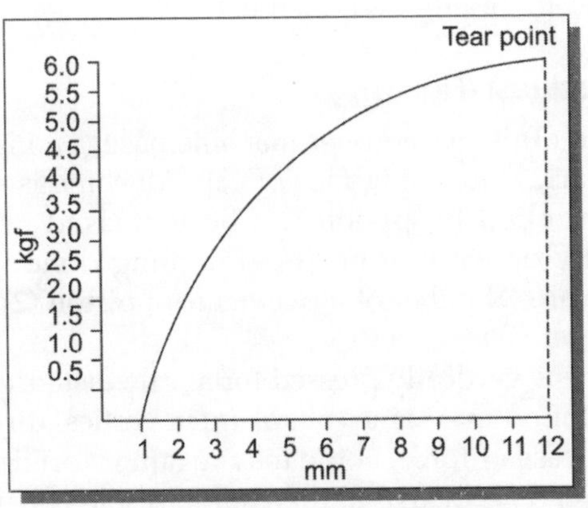

Fig. 8.13: Force versus distance

TESTING OF BOTTLES

Plastic bottles are produced from polymers such as polyethylene, high density polyethylene, polyethylene terephthalate (PET), PVC, etc.

These bottles are used for packaging drinking water, soft drinks, beverages, edible oils, wines, alcohol, ghee, etc. Objectives of various tests are to ensure that bottles do not get damaged during transport, there is no adverse interaction between contents and plastics material of bottles.

Following standard tests are generally performed:

ASTM 2463-10b	Standard test method for drop impact resistance of blow-molded thermoplastics containers
ASTM D 1434 and ISO 26	Gas permeability test
ASTM D 2063, 3198	PET bottle cap Torque test
IS 2798:1998	Methods of test for plastics containers (first revision)
IS 6312:1994	Polyethylene containers for transport of materials
IS 14764:2000	PET containers for packaging ghee –specifications

It is not possible to reproduce test methods standardization by BIS and ASTM or any other test method. However, few typical test methods are briefly described here.

Vertical Load Test (Fig. 8.14)

Caped bottle (3) is placed over machine base (1). Piston (2) of a pneumatic jack is touching face of cap. Machine is so designed that force applied by piston can be increased or decreased automatically depending on programming done while using testing software. Number of movements of piston (2) per minute may also be set.

If the bottle is vertically pressed for a little distance, say 5 mm, it would be found that some collapsing in vertical direction takes place. On release of force, bottle may regain its original shape. If force of piston (2) is increased then bottle would collapse more in comparison to first time. On release of forces bottle may or may not attain its normal shape. A situation may arrive when bottle is collapsed to such an extent that it does not attain its normal shape back. So the maximum force by which shape of bottle is not permanently de-shaped is the safe force which can be applied. But it may not be a final result. If same load is applied in cyclic manner (number of pressings per minutes) then the result may be different. So prescribed cycle and steps of load increase may be set. Visual observations are done to see that at what point of settings partial deformation remains.

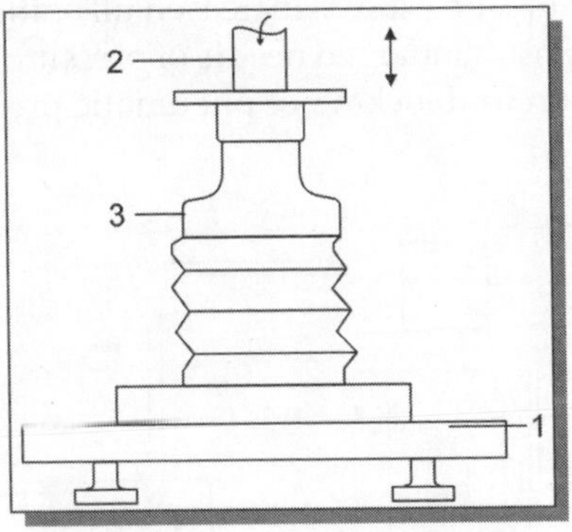

Fig. 8.14: Vertical load tester

Drop Test

Bottle filled with contents (water, other liquid, powder or granules) and capped tightly are dropped from pre decided height such as 1, 1.5, 2, 2.5, and 3 metres, etc. Due to impact, bottle may crack, burst or cap slips out. In such a case, drop height may be reduced. In fact testing starts from less height and gradually increased till bottle fails.

Bursting Test

During handling and transport of bottles, filled with liquid may get subjected to pressing on diameter. Consequently, a hydraulic and/or pneumatic pressure gets develop inside the bottle. If walls of bottles are strong then bottle will not burst, otherwise it will burst. At the same time fitting of caps should be so strong that it should not slip out of bottle threads in the event of development of bursting pressure inside the bottle.

Figure 8.15 shows basic construction of a typical burst testing machine/instrument. (1) is pneumatic pressure adjustment and indication system. (2) is an enclosure with polycarbonate window for observation. (3) is a water reservoir. Pneumatic pressure is applied on the top surface of water in the reservoir. Bottle to be tested is filled with water and screwed in a holder which is located inside the safety enclosure and connected to reservoir through suitable pipes. (5) is a pneumatic pressure indicator. There is a red color needle which advances with indicator needle which is generally black in color. (7) are the knobs to increase or decrease pneumatic pressure. Pressure is increased till bottle burst. Bursting pressure may be read under red needle of pressure indicator meter. There is provision for blocking of pneumatic pressure as soon as bottle bursts.

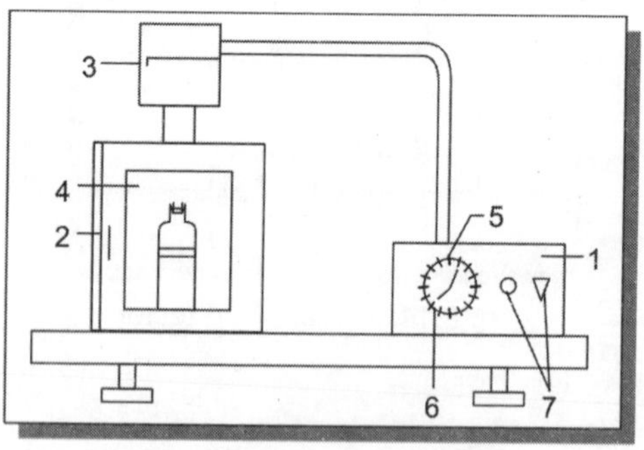

Fig. 8.15: Burst tester

TESTING OF PLASTICS PRODUCTS

Plastics products may be article which itself are products such as pipes, sheets, mugs, trays, plates, tapes, ropes, woven sacks and a host of such items.

There are innumerable electrical, mechanical and electronic products in which some plastics components are used. Makers of high value products generally use plastics parts which are tested by producers, according to standard test methods and test results are made known to purchaser through brochures , written or verbal declaration.

Selection of test mostly depends on design and end use of component.

A thermoset handle fitted on a kitchen utensil is likely to be subjected to impact and bending action. Handle should be strong enough to withstand impacts and bending actions. So standard tests for tensile, bending and hardness may be carried out to ensure that utensils handle will not fail during normal service.

TEST SPECIMEN PREPARATION

Preparation of test specimen is a critical step in testing process for a polymer testing. Plastics may be in the form of granules or sheets. Hence following two methods are generally adopted for preparing specimen.

1. By injection molding
2. By machining

In both the cases, specimen must conform to shape, dimension, finish as specified by standard test method. This essential requirement is further explained in following manner.

Let there be two numbers of specimen of same material for testing izod impact test having dimensions.

Specimen no 1 has a width of 12.0 mm and thickness 2.5 mm. Specimen no 2 has a width of 12.1 mm and thickness 2.6 mm.This difference in dimension is not intentional. It is due to machining or blanking inaccuracy. Since cross-section area of specimen no 1 is less as compared to specimen no. 2, test result will indicate different values of impact strength. But it is not true as specimens are prepared from same material. Results are misleading due to variation in dimensions of test specimen.

By Injection Molding

There are standard methods to carry out molding with special mold design (mold design may be as per ISO 294 and ASTM D 3641), cavity dimension and molding parameters. Object is that specimen should be free from internal stresses and other molding defects. Dimensions of specimen after cooling down to room temperature and seasoning should be within a tolerance of + or – 0.015 mm or as specified by test method.

Shapes of specimen may be like the ones shown in Fig. 8.16.

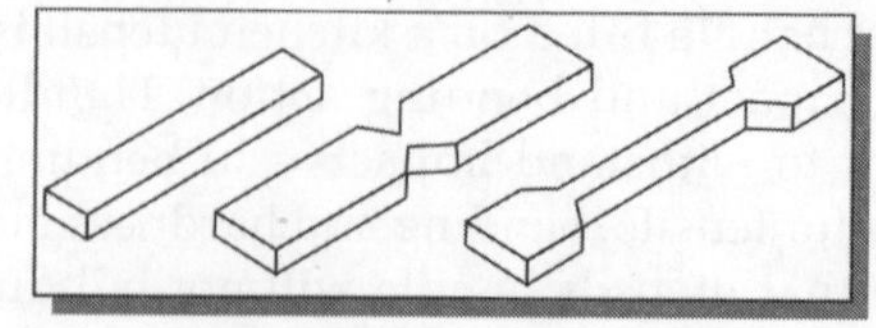

Fig. 8.16: More specimen

By Machining

In machining a specimen from a sheet or solid piece, internal or *under surface* stresses are developed, if machining is heavy and fast. Precision table mounted machines are developed to carry out machining operation such as cutting, turning, milling, filling, polishing, etc. Machining accuracies of + or – 0.010 may be achieved. In case of preparing a test specimen from a sheet or film, precision made blanking and cutting dies are also used. These tools (punch and die) are fitted on a precision small press to carry out blanking operation.

Due to copyright restriction, IS, ISO or ASTM specimen test methods cannot be reproduced here. It is recommended that these standards may be consulted to enhance knowledge. It is further recommended that following websites may be visited

www.ptli.com
www.testopedia

TEST

1. Why is it necessary to have standard test methods for testing plastics materials and products?
2. Name few national and international test methods.
3. Name few tests which are mostly performed at R & D, factories and by end users.
4. Write down website addresses from where information regarding tests and testing equipments may be retrieved.
5. What are the differences between the tests for polymers and polymer products?
6. Prepare a table to provide following information, S no, test methods (at least five), units, values for HDPE, polycarbonate and ABS.

7. Categorise tests into following:
   - Mechanical
   - Electrical
   - Optical
   - Chemical
   - Thermal
8. With what objectives, producers of polymers, test laboratories and bulk producers of products perform various tests?
9. Describe with the help of sketches following testing equipments:
   - Universal testing machine
   - Flexural testing machine
   - Melt flow rater
   - Vicat softening point
10. Describe with the help of sketch method of using following testing equipment:
    - Izod impact tester
    - Hardness tester
    - Abrasion tester

ASSIGNMENT

Note: Tick mark appropriate options.

1. Tensile strength of a polyamide strand of standard diameter is tested by
   (a) Vicat softening tester .... ☐
   (b) Density testing equipment .... ☐
   (c) Universal testing machine .... ☐
2. Use of standard test method is desirable due to which of the following reasons
   (a) Not expensive .... ☐
   (b) To convey feeling of reliability .... ☐
   (c) To make sure that plastics of desired quality is being selected .... ☐
3. Tick mark values of melt flow index which is nearest to correct value HDPE molding grade

(a) 5 to 10 gm per 10 minutes ☐
(b) 25 to 30 gm per 10 minutes ☐
(c) 5.2 to 8.5 gm per 10 minutes ☐

HDPE extrusion grade

(a) 0.5 to 2 gm per 10 minutes ☐
(b) 2.1 to 4 gm per 10 minutes ☐
(c) 5 to 10 gm per 10 minutes ☐

4. Tick mark values of tensile strength at break which is nearest to correct value polypropylene?
(a) 5000 psi ☐
(b) 20,000 psi ☐
(c) 8000 psi ☐

Polypropylene 30% fiber glass filled

(a) 8000 psi ☐
(b) 10,000 psi ☐
(c) 5000 psi ☐

5. Which of the following polymers has maximum percentage of light transmittance?
(a) GP polystyrene ☐
(b) PET ☐
(c) PMMA ☐
(d) Polycarbonate ☐

6. Which of the following dimensions of specimen for testing izod impact test is correct?

	L × W × T		
(a)	73.5	13.5	3.5 ☐
(b)	63.5	12.5	2.5 ☐
(c)	53	10.2	4.1 ☐

7. In tensile testing, sometimes strain gauge is used. Which of the following purposes is correct?
(a) To measure diameter of specimen ☐
(b) To measure pulling force applied ☐
(c) To measure increase in length of a particular portion of specimen ☐

8. Bursting test on bottles is performed in which of the following manners?

- (a) By applying hydraulic pressure .... ☐
- (b) By applying gas pressure while bottle is placed on a table .... ☐
- (c) After placing bottle in an enclosure .... ☐

9. In which of the following manners pipes are tested?
   - (a) By closing one end and attaching other end to water tap of OH tank .... ☐
   - (b) Closing one end and applying pneumatic pressure .... ☐
   - (c) By using a jig and hydro-pneumatic machine .... ☐

10. Modern testing equipment are generally fitted with which of the following?
    - (a) PID .... ☐
    - (b) PLC .... ☐
    - (c) Microprocessor with display screen .... ☐

Index

Abrasion test 260
Abrasion tester 266
ABS 8
Acceptance sampling 208
Acetal 9
Additives 24, 48
Addresses 260
Air vent 89
All hydraulic machine 46
Angle of incidence 149
Angle of refraction 149
Annular blown film 104
Appliances 11
ASTM 10
ASTM tests 145
Automatic 182

Band heaters 52
BIS 10
Blow molding 117
Blow pin 251
Blown film 99
Bottle testing standards 276
Break even point 229
Bubble section 105
Burning test 260

Calendaring 122
Carriage 251
Cast films 99
Cavities layout 56
Cavity filling pattern 88
Celluloid 6
Choking bar 101
Coefficient of friction tester 270
Co-extrusion 99
Cold well 73
Collapsible core 84
Combination machines 37
Communication ways 234
Components 86
Composite molding 126,128
Compounding 99
Compression molding 23
Contingency funds 227
Convertible machines 33
Covalent bond 1
Covalent compound 2
Cylinder nozzle 53

Defects in molded components 86
Degating 58
Deviation analysis 205
Die lip 102
Dielectric constant 151
Digital display screen 248

Ejection system 45, 82
Elastic limit 367
Electro-hydraulic 133
Electro-mechanical 133
Electrovalence 2
Embossed surface 102
Engineering plastics 16
Epoxy 19, 20
Etching of surface 255

Extruder screen 98
Extrusion 94
Extrusion die 103

Fail safe design 41, 82
Fan gate 63
Fillers 19
Fits and tolerances 160
Fixed and variable cost 192, 223
Fixed cost factors 223
Fixed mold platen 42
Flexural modulus 142
Flexural strength 260
Flow back 62
Flow of plastics 69
FRP 125, 126
Functional tolerance 7

Gas assisted molding 89
Gates, shapes 63
Gates, types 60, 64
Granules transport 50

Hardness 260
Haze 149
Heat transfer 70
Hopper loading 49
Horizontal injection molding machine 35
Hot runner system 59
Hydraulic elements 166
Hydraulic power 156
Hydraulic principles 164

Inching 181
Injection blow molding 121
Input components 199
Insert molding 33
Intermittent extrusion blow molding 119
Ionic compound 2
IS test website 262
ISO standards 209
Izod impact 269
Izod impact tester 265

LCP 18
Light transmittance 13, 149
Locking force 46

Machine units 31
Machine, horizontal 31
Machines categories 31
Maintenance categories 161
Maintenance management inputs 189
Management over head cost 225
Manifold 60
Manuals 189
Marginal costs 229
Matching feedhead 53
Mechanical advantage 165
Mechanical properties 141
Melamine 19, 22
Melt flow index 148, 260
Melt flow rater 262
Method, surface treat 254
Methodologies, blow 118
Microprocessor 243, 247
Modifiers 48
Mold closing and opening 79
Mold temperature 78
Molding sequence 119
Molding cycle sequence 81
Molding machine 30, 35
Moving injection unit 43
Moving mold plate 42
Multicolor molding 75
Multi-core operation 80
Multi layer sheet 107
Multi-core mold 80

Name of few standards 260
Nanomaterials 26
Nanoparticle size 26
Non-drool nozzle 56
Nozzle design 55
Nucleus 2

Octet rule 2
Optical properties 148
Organizational structure 235

Packing in cavity 62
Parameters, process 244
Parameters, machine design 30
Processing parameters 30
Parison 117, 118
Pellets 47
Percentage of fillers 148
Performance 147
Performance plastics 145
Permeability 260
PET 17
Phenol formaldehyde 18,20
Phenolic 19
PID 186
Pipe testing factors 271
Pipes 99, 272
Plant and machineries capabilities 198
Plasticization 50
Plastics flow path 52
Plastics granules and pellets 47
Plastics waste 129
PLC 79
PMMA 13, 16
Polyamide 9, 17, 18
Polyblend 24
Polycarbonate 9, 16
Polyester 19
Polyester thermoplastics 10
Polyethylene 10
Polymer 1
Polypropylene 16
Polystyrene 16
Prepare process cost 222
Pressure intensifier 47
Principles of costing 219
Process parameters 200
Preparation processing cost 222
Processing temperature 148
Product categories 154
Productivity 187, 188
Profile extrusion 108
Profile parison wall thickness 120
Properties classification 8
PVC 17

Quality assurance 198
Quality management organization 139
Quality manual 210
Quality testing in house 155
Quality uniformity 73

Refractive index 149
Relief valve varieties 170
Rubber 1
Runner layout 56
Runners 56

Safety precautions 93
Scrap 7
Screen assembly 98
Screw design 96
Screw zones 96
Semi automatic 182
Sequence of operation (blow mold) 119
Setting processing parameters 244
Sheet draw behavior 115
Sheet lamination 125
Sheets 99, 101
Shot weight, shot volume 36
Specific gravity 15, 148
Specific gravity testing 264
Specimen preparation 280
Sprue 32
Sprueless nozzle design 55
Standard and marginal costs 221
Statistical sampling 208
Steel specification sheet 41
Stretch blow molding 122
Structure, organize 235
Submarine gate 66
Surface treatment 254

Tab submarine gate 67
Take off equipment 101
Tear strength 260
Tensile strength 260
Test houses 154
Test methods 140
Testing agencies, web addresses 260
Thermal properties 144
Thermoforming 109
Thermoplactics names 7
Thermoplastics 1
Thermosetting 1
Three plate mold 59, 65
Tie bar steel 40
Tie bars 38
Time dependent performance 147
Toggle joint 44
Transmittance of ultra rays 149
Tubing 99
Types of organizations 236
Typical values of fits 159

Universal testing machine 276
Unscrewing system 84
Ultraviolet rays, trans 149

Vacuum forming machine 113
Vacuum forming system 117
Vacuum sizing fixture 106
Vertical injection molding machine 31
Vicat softening point 260

Wastage elimination 54
Waste generation 130
Waste management 131
Weather effect 260
Websites for testing pipes 272
Wire covering 99

Reader's Note

Reader's Note